Plants in the Civil War

Also by Judith Sumner

Plants Go to War: A Botanical History of World War II (McFarland, 2019)

Plants in the Civil War

A Botanical History

JUDITH SUMNER

McFarland & Company, Inc., Publishers
Jefferson, North Carolina

This book has undergone peer review.

Library of Congress Cataloguing-in-Publication Data

Names: Sumner, Judith, author.
Title: Plants in the Civil War : a botanical history / Judith Sumner.
Description: Jefferson, North Carolina : McFarland & Company, Inc., Publishers, 2022 | Includes bibliographical references and index.
Identifiers: LCCN 2022038610 | ISBN 9781476691312 (paperback : acid free paper) ∞ ISBN 9781476648835 (ebook)
Subjects: LCSH: Botany, Economic—Southern States—History—19th century. | Plants and civilization—Southern States—History—19th century. | Plants, Useful—Southern States—History—19th century. | Plants, Cultivated—Southern States—History—19th century. | BISAC: SCIENCE / Life Sciences / Botany | HISTORY / Military / United States
Classification: LCC SB108.U6 S86 2022 | DDC 581.60975—dc23/eng/20220831
LC record available at https://lccn.loc.gov/2022038610

British Library cataloguing data are available

ISBN (print) 978-1-4766-9131-2
ISBN (ebook) 978-1-4766-4883-5

Front cover: *insets, left to right:* illustrations of cotton plant, tobacco plant and asafetida plant (Adobe Stock); *background:* idealized illustration of southern plantation (Shutterstock/Everett Collection)

Printed in the United States of America

McFarland & Company, Inc., Publishers
Box 611, Jefferson, North Carolina 28640
www.mcfarlandpub.com

Table of Contents

Preface

The Civil War was like no other time in American history, a prolonged internal conflict provoked by the cash crops that drove the Southern economy. Enabled by enslavement, the Confederacy fought for the right to cultivate cotton, sugar, tobacco, and rice on a vast scale. As the ethnobotanist Richard Evans Schultes observed on occasion, historians stumble because they know little about botany; according to his worldview, knowledge of plant uses is fundamental to understanding civilization and interpreting conflict. Wartime needs for food, drugs, fibers, and timber in both the North and South relied on the botanical properties of diverse plants, some used traditionally and others adopted in wartime. From the military perspective alone, plants provided virtually all of the natural products essential to the Civil War effort, from fibers for uniforms and timber for military engineering to dietary rations and antimalarial drugs.

This plant-centered history of the Civil War examines in detail the economic botany and ethnobotany of the Civil War period, including discussions of civilian foodways and military diet; medicinal plants and practices; cultivation and management of cash crops; plantation landscapes, farming practices and gardens; and the ethnobotany of enslaved people. My goal in writing this book has been to provide a synthesis of various plant uses and botanical connections as they relate to the American Civil War. It is another contribution to Schultes' desire for history viewed and interpreted through a botanical lens, to some extent a companion volume to my earlier book *Plants Go to War: A Botanical History of World War II* (McFarland, 2019).

Wartime is typically a period of innovation sparked by need. This project has intentionally emphasized civilian and military life in Confederate states, which experienced a greater impact from wartime conditions than the North. Many botanical products and commodities were in short supply as a result of the Union blockade, and the devastating effects of war on farms and land resulted in additional hardships. In *Resources of the Southern Fields and Forests* (1863), Frances Peyre Porcher compiled a vast array of known plant uses and potential botanical substitutes, and this wartime compendium served as a valuable guide in understanding both civilian and military needs. Porcher described and suggested possible foods, medicinal plants, dyes, textile fibers, resins, oils, timber, cordage, insecticides, soaps, waxes, tonics and bitters that could be harvested or produced from local plant populations. Many Southerners returned to traditional uses or experimented with alternatives to traditional foods, fibers, medicines, and other botanical commodities, some gleaned from Native American ethnobotany. Southern biodiversity, landscapes, plantations, and military action all informed the narrative, as did the ethnobotany of the enslaved people whose labor fueled the Confederate economy.

Many pertinent nineteenth-century resources are now extensively digitized, which suited this project for prolonged social isolation during the recent coronavirus pandemic. I am particularly grateful to the Biodiversity Heritage Library, the Hathitrust Digital Library, the Internet Archive, and to the libraries at Princeton, Duke, and Cornell universities for digitizing and sharing numerous useful books and periodicals, including the Southern agricultural journals that provided insights into Civil War stringencies and adaptations. My research involved a wide range of primary sources, including newspapers; journals and magazines; cookbooks; household, farm, and agriculture manuals; military manuals and texts; period agricultural, gardening, and medical texts; travel reminiscences; and the slave narratives recorded by the Works Progress Administration during the 1930s. The latter have been criticized for providing a sanitized, possibly simplistic view of slavery; although over 2,200 interviews were conducted in 17 states, there was no particular plan to interview a cross-section of former slaves. However, many individual accounts recollect the practices and self-sufficiency of enslaved people during the antebellum and war years. There is no particular reason to suspect that these anecdotal memories are inaccurate, although in some cases vernacular plant names are obscure or inconsistent.

Once again I am grateful to family members for their interest and encouragement. My husband, Stephen Sumner, has served as a sounding board as the project unfolded and the narrative of Civil War botany took shape. His vast knowledge of military history helped to contextualize much of what I discovered in the literature. Our daughter and son-in-law, Dr. Catherine Sumner and Dr. Stephen Eyre, answered medical queries with dispatch and provided encouragement during the somewhat dismal days of social isolation and prolonged winter weather. Our grandchildren, Jeremy and Lillian Eyre, brightened the horizon with their good cheer, much of it shared over FaceTime. Once again I hope that this book will provide insights for readers with interests in history, horticulture, and plant science, in particular the economic botany and ethnobotany of the Civil War era. Given the scope of the project, I regret that omissions seem inevitable, and I am solely responsible for any errors. Information on plant uses during the Civil War is shared for historic perspective only and not to suggest or recommend medical treatments, herb uses, or experimentation with any medicinal species mentioned in the text.

1

The Botanical Roots of Slavery

Colonial Origins

Slavery in the Confederate states evolved from enslavement of Africans in the original thirteen colonies, which included Massachusetts, Connecticut, Maryland, New York, and New Jersey. The Body of Liberties (1641), the legal code of the Massachusetts Bay Colony, allowed human bondage, and for the next 130 years, most colonial exports in some way depended on enslaved labor for their production. Enslaved people in Northern colonies became skilled carpenters, blacksmiths, printers, and coopers, and they were involved in building the ships that carried raw materials to the Caribbean, where planters grew the sugar cane needed to make rum. In turn, rum was traded for Africans who were needed to cultivate cane sugar for rum production, and profits were used to buy Caribbean sugar (often in the form of molasses) for the New England distilleries that made the rum.

The North depended upon skilled trades as the basis of its colonial economy. In New England some enslaved people worked in fields, but most learned shipbuilding, printing, carpentry, and other trades. This contrasted dramatically with the South, where the economy pivoted on highly productive crops cultivated on a vast scale by enslaved workers. Indigo, rice, tobacco, sugar, and cotton are tropical or semi-tropical plants that thrive in the Southern climate, and they yielded vast quantities of high-demand nonperishable products which were easily exported. Most cash crops required some sort of processing prior to sale, which made enslaved workers an even greater perceived need in the plantation economy.

The dye plants known as indigo (*Indigofera tinctoria* from Asia and *I. suffruticosa* from tropical America) were among the earliest crops planted by colonists. As plants that yield stable, deep blue hues, indigo was valued as the source of a dye for cloth. Europeans used indigo imported from India during the sixteenth century and then tried cultivating the plants in tropical or semi-tropical colonies rather than relying on imports. There were large slave-cultivated indigo plantations in colonial South Carolina, but these were abandoned when the British market for American indigo waned during the Revolutionary War.

Tobacco (*Nicotiana tabacum*, native to tropical America) was a significant colonial cash crop in the both the North and the South, but cultivation and curing were labor intensive. By 1620, Virginia growers were auctioning crops for export to Europe, where herbalists recommended tobacco for its purported medicinal properties, including the notion that it prevented plague. Enslaved Africans were employed in tobacco production, which included the clearing of new land because consecutive crops quickly depleted soil nutrients. By the start of the Civil War, over a third of a million enslaved people labored in tobacco fields and barns, which resulted in a vicious cycle of production and enslavement. The more land that was cleared, the more tobacco crops were sown, and the more enslaved workers were needed for cultivation, harvesting, and curing.

Rice (*Oryza sativa*) is an Asian grass that requires standing water, and the mud-walled paddies of traditional Asian agriculture were simulated by the tidal swamps and river flood plains of South Carolina. Enslaved Africans cleared native vegetation and built levees to maintain the flooded rice fields. By the 1730s, Charleston growers sold over 20 million pounds of rice annually. Imports soared in the years leading up to the Civil War, and rice became another crop that without enslaved workers would have been impossible to plant, cultivate, and harvest.

Colonial plantations included crops of Asian indigo, a legume that yielded a stable deep blue dye from its leaves. Early crops were exported to England before the Revolutionary War (author's collection).

Rice crops thrived in the tidal swamps and flood plains of South Carolina, where they became a lucrative colonial export. Enslaved workers built the levee system that flooded rice fields to resemble Asian paddies (Harter, 1998).

Sugar cane (*Saccharum officinarum*) originated as a wild grass in Papua

New Guinea, and the natural canes were introduced worldwide as an edible delicacy. Growers in India developed the first method for refining and crystalizing cane juice into sugar, a commodity that arrived in Europe by the Renaissance. Columbus introduced stem cuttings to Haiti, and Spanish settlers planted cane fields in the Canary Islands and Caribbean, where African enslavement became linked with sugar cane cultivation. By 1850, American growers had established sugar cane plantations with hundreds of cultivated acres in Maryland, North Carolina, South Carolina, and Georgia. Crops were grown as a perennial and harvested up to four times before fields were burned and replanted. All stages of cane planting and sugar production depended on the labor of enslaved workers and dwindled after Abolition.

Fibers harvested from the capsules of cotton (*Gossypium hirsutum*) became the economic "King Cotton" of the American Southern colonies. Colonial Virginians grew cotton as a cash crop, and by the early 1700s it was widely grown across the South. The invention of the cotton gin in 1792 transformed the tedious task of separating the fibers from the seeds, and the demand for fibers increased as mills appeared across New England. Enslaved labor was exploited to meet the demand for cotton, which during the 1800s became the main crop of most Southern plantations and the economic fuel of the Confederacy.

Enslaved Life

During the 1770s, New England states began to abolish slavery, and by 1804, legislation ending slavery was in place in every state in the Northern United States. Thus nineteenth-century slavery was limited to the South, where about one third of the population were Africans and their descendants, comprising about four million enslaved people. They lived close to the land, depending largely on vegetable diets, living in timber cabins, and planting, cultivating, and harvesting crops. Proximity fostered an environment in which useful plants were shared and exchanged, and practical information about the propagation and uses of cultivated and native plants became part of their oral history.

The shift from indentured servants to enslaved workers affected the productivity of Southern agriculture at the expense of human lives. Enslaved existence was far from idyllic; they worked longer hours and more days than servants, and women and children were required to join the men in the fields. Often interpreted as resulting from improvements in machinery and techniques, increased agricultural yield in Southern colonies in fact resulted from the shift to enslaved labor.[1] Slavery-fueled rice cultivation in the Carolinas provided the social and economic model for the cotton plantations that followed, which also depended on slavery to realize the goals of large scale plantation agriculture.

There were similarities shared by colonial enslaved people regardless of region, including the expectation of self-sufficiency. Living quarters in the colonial South differed from the idealized timber cabins built to calm Abolitionists' outrage and illustrated in nineteenth-century publications. Early dwellings were patterned on

African buildings and assembled from wattle. The technique relied on poles set in trenches and interwoven with grasses and woody branches, sealed and reinforced with layers of mud with a high clay content. Roofs were thatched using the leaves of palmetto palms; openings in the wall served as windows, and fireplaces were generally lacking. Dwellings were the size of single rooms, no larger than about twenty feet long and fifteen feet wide, and sheltered one or more families. Furnishings were few or lacking. Outdoor pits were used for cooking fires, and corn cobs were often burned nearby to rid dooryards of insects.

Some growers may have objected to having thatched huts on their land, or perhaps more conventional housing evolved from economic concern for workers' health and reproduction. Improved nineteenth-century dwellings were often cabins with

Nineteenth-century plantations included primitive one-room cabins for enslaved workers and their families, in stark contrast to antebellum plantation houses. Abolitionists forced many planters to improve living conditions by replacing thatched huts with dwellings made of logs or lumber (Charles C. Coffin, *Building the Nation* [New York, 1883]).

log foundations and walls, sealed with mud and straw against the elements, and roofed with wooden shingles or boards. Fireplaces and chimneys were molded from mud and sticks, and some dwellings had floors made of wooden boards, an improvement over the earthen floors of thatched huts. Cabins were also constructed using wooden frames clad with boards that more closely resembled the houses of planters. On many plantations, decay-resistant bald cypress was the wood of choice for building cabins for enslaved workers (see Chapter 8).

Larger plantations in the South had dozens of such cabins, whitewashed or left to weather, constructed in rows along a central road and surrounded by gardens of vegetables and medicinal herbs. However, only 12 percent of all owners had twenty or more enslaved people to provide for, so such landscapes were relatively rare. Individual accounts described "cabins as far as you could see" with some dwellings situated near the agricultural fields and others built near the main house, depending on the labor required of the enslaved workers.[2] Wooden furnishings included box-like bedsteads filled with pine needles or beds with mattresses filled with straw, cornhusks, or moss. By providing improved housing for families, overseers encouraged the birth and rearing of more enslaved workers. Structures were nevertheless built to last no more than a few decades and were likely to succumb to fungal decay and weathering.

Moreover, designs merged and depended upon the available materials and local customs. At a time when land was still being cleared for farming, some masters and overseers also occupied log cabins, albeit larger and surely better furnished than workers' dwellings. Planters had elaborate homes with architecture that was sometimes mirrored in the construction of adjacent wood frame quarters built for enslaved people who worked indoors. These differed from typical cabins in having masonry foundations. By the mid-nineteenth century, Abolitionist criticism and reform movements resulted in improved housing for some enslaved people, with changes that included glazed windows, wooden shutters, raised wooden floors, and masonry fireplaces.

Independence among enslaved people was generally discouraged, particularly if it might result in financial gain. While they were expected to plant vegetable gardens for their own use, many owners forbid enslaved workers from cultivating any crops (including cotton) that could be sold. The outcome was minimal awareness among enslaved people of marketplace economy and trade. Owners preferred to provide a cash allowance rather than risking the resourcefulness or rebellion that financial independence might foster among workers.[3]

Economy and Natural History

Southern devotion to an agrarian way of life depended on slavery for labor, especially on farms exceeding one hundred acres. Crops dictated the extent and nature of farm work and the number of enslaved people needed for planting, cultivation, and harvesting. A diet of vegetables and bread, supplemented by a modest meat allowance, kept enslaved people healthy enough to work. During the 1860s, their labor fueled the Confederate war effort by providing food and cotton clothing for Confederate troops. Enslaved people also constructed fortifications and built railroads, and some served as cooks, musicians, or laborers in the Confederate Army.

Their labor became an economic necessity on large farms and plantations, yet only about a quarter of all Southerners owned enslaved workers, and of these, half owned five or fewer. Most farmers cultivated fewer than a hundred acres, but many were aspirational and envisioned large scale agriculture. Although the owners of large plantations comprised less than 0.5 percent of the population, the planter class provided Confederate leadership. Nevertheless the acceptance of slavery was widespread and rationalized based on interpretations of morality and nature. The conservative Southern view was that moral character was an intrinsic fiber of the agrarian economy and that enslavement provided shelter, care, and training for people who could not fend for themselves.[4]

By the 1850s, some justification for slavery arose from biology, for instance the interpretation of human biology argued by the Harvard paleontologist and geologist Louis Agassiz. Along with rejecting the idea of biological evolution by natural selection, Agassiz interpreted variation among *Homo sapiens* as the work of a creator, evidence for the existence of several races with different origins and distinct traits. He

Enslavement had colonial origins because tobacco and other cash crops required a large work force, and plantation agriculture was unsustainable without free labor. Before the Civil War, about a quarter of all white Southerners owned enslaved workers ("A Tobacco Plantation," published by Bowles and Carver, New York Public Library).

held that races were designed for particular "provinces," a belief system that he popularized as the theory of polygenism.

Proponents embraced the idea of a separate black race (sometimes described as a developmental midpoint between humans and other primates) as a rationale for slavery, which became the basis for scientific racism. Drawing on polygenetic thinking, in *Negroes and Negro Slavery* (1853 and later editions) John Van Evrie defended slavery by arguing that Africans were inherently inferior. He maintained that slavery was Africans' "normal condition" and that education could actually harm their physiology. Perhaps ironically, current research suggests Africa as the cradle of human evolution, where early *Homo sapiens* first distinguished themselves as a species unique from other primates. It appears we are all descended from early humans who became adapted to life in African environments, a realization that would have been quite surprising to proponents of scientific racism as grounds for slavery.

A natural justification for slavery was embraced by some botanists as well. During the antebellum years, the South Carolina planter and mycologist Henry William Ravenel worked diligently on his collection of scientific journals and dried fungi, botanical endeavors made possible by the enslaved people who worked in his fields. He recorded in a journal after the war, "There must ... be stringent laws to control the negroes, & require them to fulfill their contracts of labor on the farm."[5] As both a naturalist and devout Episcopalian, Ravenel may also have viewed subservience as part of natural order, even after Abolition converted the role of many former enslaved people to servants and field hands. Enslaved labor had allowed him to pursue science, but Ravenel was reduced to poverty after the Civil War. He spent the last years of his life selling botanical collections rather than pursuing the taxonomy of fungi, and he is commemorated by the genus *Ravenelia*, fungal parasites that cause rust diseases on legumes.

Charles Darwin witnessed firsthand the treatment of enslaved people in Brazil, where he saw a child beaten, and he recorded in his journal the irony that men "who profess to love their neighbours as themselves" engage in the evil practice of owning and selling others. He summarized his thoughts about slavery in the Southern United States: "It makes one's blood boil, yet heart tremble, to think that we Englishmen and our American descendants, with their boastful cry of liberty, have been and are so guilty: but it is a consolation to reflect, that we at least have made a greater sacrifice, than ever made by any nation, to expiate our sin."[6]

Harvard botanist Asa Gray corresponded regularly with Darwin and kept him current on the progress of the Civil War. At the end of the war in a letter that also referenced primroses, plantains, and loosestrife, Darwin responded to Gray, "Well I suppose we shall all be proved utterly wrong who thought that you could not entirely subdue the South. One thing I have always thought is that the destruction of Slavery would be well worth a dozen years war."[7] A few years later, Charles Darwin argued his viewpoint of human evolution from a single hominid ancestor in *The Descent of Man* (1871). Although the evolutionary details have been supplanted by more recent discoveries, the notion of a single line of evolution to *Homo sapiens* remains unchallenged.

Gray was one of the first American scientists to embrace Charles Darwin's theory of evolution by natural selection and human evolution from a single ancestral species. Like Darwin, Gray vehemently opposed slavery and regretted that he did not have a son who could enlist in the Union army. The Civil War disrupted his research in North America flora because plant specimens from Confederate states became difficult to obtain as Southern botanists became isolated from colleagues in the North. The first edition of Gray's well known botanical manual (*A Manual of the Botany of the Northern United States: from New England to Wisconsin and South to Ohio and Pennsylvania*) was published in 1848,[8] but he continued to work on revisions and new additions during the war years. His final edition (1867) was published with a revised title (*Manual of the Botany of the Northern United States, Including the District East of the Mississippi and North of North Carolina and Tennessee*) which reflected the inclusion of more Southern species after the Civil War.[9]

Popular Culture

Enslavement became symbolic of the Southern way of life, often falsely cast in an idyllic agrarian light. Idealized media images of the pro-slavery rural South proliferated as sentimental mainstays of the nineteenth century, even in song. Although he had not seen the Confederate states firsthand, Stephen Foster conjured the "little farm" and "one little hut among the bushes" with "bees a humming, all 'round the comb" in the lyrics of "The Old Folks at Home" (1851). "Old Boys, Carry Me 'Long" (1851) and "Old Black Joe" (1853) mentioned fields of sugar cane, tobacco, and cotton, with references to enslaved labor. "My Old Kentucky Home" (1853) idealized farm life in which "the corn top's ripe and the meadow's in the bloom" and "the young folks roll all around the cabin floor, they're merry, all happy and bright," perhaps inspired by imagery from Harriet Beecher Stowe's *Uncle Tom's Cabin* (1852). Stowe also had not visited the South but understood intuitively that slavery worked against the natural order of independence, parity, and family life.

Currier and Ives, New York, lithographers, billed their firm as "the Grand Central Depot for Cheap and Popular Prints" that provided "colored engravings for the people." Their business centered on familiar, intriguing, sentimental, rural, and patriotic themes, with hand-coloring done by an assembly line of German immigrants. They produced a cluster of chromolithographs that suggested the peaceful co-existence of owners and enslaved people, but there were not published until after the Civil War. In some ways, these lithographs spread news of current events, and they presented an idealized version of Reconstruction. "Loading Cotton on the Mississippi" (1870) depicted workers loading baled cotton onto the deck of a steamboat, with African youth playing carefree in the foreground, against the natural beauty of a forested bluff.

Commissioned by the federal government and based on an illustration of an antebellum plantation in Louisiana, the Currier and Ives chromolithograph "A Home on the Mississippi" (1871) illustrated a riverfront home in a grove of hardwood trees bearing Spanish moss. Emancipated slaves are seen riding in a carriage, walking along the river and by the fence, suggesting a camaraderie and social order accepted by all. Over a decade later, "A Cotton Plantation on the Mississippi" (1884) illustrated former enslaved people picking cotton. In the foreground, there is a well-dressed couple, very likely the planter and his wife, in dramatic contrast to African couples in the field. In the background, there is a cabin and a woodland in which several of the trees appear dead; whether this is meant to suggest the war-damaged Reconstruction landscape and demise of slavery is unclear, but the contrast between the working and leisure social classes remains stark.

Currier and Ives chromolithographs recorded social interactions and class against a background of natural beauty, orderly agriculture, and botanical diversity. These intentionally refined images differ drastically from the "Darktown Comics" published by Currier and Ives during the same years. Over 73,000 of this series of lithographs were sold between 1879 and 1890, featuring images that clearly reflected the use of scientific racism as a justification for slavery. The drawings mocked the incompetence and foolishness of former enslaved people and bore legends written in dialect. They continued to support the notion of black subservience as part of the natural order, even after Emancipation.

2

Plantation Landscapes

Land and Trees

As farming expanded, plantations were carved from ancient primary growth forests across the South. By the antebellum period much of the wilderness had been tamed to reflect English traditions, which included houses and gardens typical of landed gentry during the seventeenth and eighteenth centuries. Harvested timber was used to construct planters' houses, cabins, churches, school houses, smoke houses, barns, and other outbuildings. The large house occupied by the planter and his family was the architectural centerpiece, situated in a landscape of contrasts amidst ornamental gardens, cultivated fields, work yards, workers' quarters, and miscellaneous outbuildings of the plantation business.

Plantations typically evolved from family farms, as land and money accumulated over years and generations. At their antebellum height, plantations functioned as virtual villages that were the antithesis of subsistence homesteads, operating instead as agricultural factories fueled by enslaved labor. Plantations produced a handful of cash crops, including vast quantities of cotton, rice, sugar, and tobacco, which were then sold at market for considerable profit. Indeed these tropical and semi-tropical crops flourished to an extent not possible in cooler states. The synergistic combination of a mild temperate climate, high rainfall, long growing season, and fertile soils created ideal conditions for cash crops grown on a vast scale.

Well-tended plantations included cultivated groves of trees that self-seeded into woodlands used for aesthetic appeal, timber crops, and climate control. Clusters of trees were often intentionally oriented to overarch and shade roads and buildings, including homesteads and working kitchens. Shade trees were considered critical for protecting the health of planters' families during the summer, a season of extreme heat and infectious diseases that included malaria. Some believed that trees created breezes and improved stagnant air; at the least, foliage offered shelter from midday sun and helped to hide necessary outbuildings and work yards.

Plantation groves included both native and non-native species, and these were selected for growth form, size, and canopy cover rather than geographic origin. Some individual trees were preserved from the virgin forests that originally blanketed plantation landscapes; others were transplanted as seedlings or selected from nursery stock of native and non-native woody plants. In his *Treatise on the Theory and Practice of Landscape Gardening* (1841), Andrew Jackson Downing observed that "these places owe almost their entire beauty to nature, as nearly all the fine trees, groves, and woods, are the natural growth of the soil ... have fallen as prey to the licentious axe of the woodman" and argued that "the modern style of landscape gardening is produced, mainly by retaining and preserving the materials of which nature has here been so extremely prodigal."[1] Even as land was aggressively cleared as agricultural soils lost nutrients, groves of trees were preserved and planted as a key element of plantation landscapes.

Style was a consideration among well-to-do Southerners who followed and favored British culture. Indeed Downing advocated for the wooded landscapes resembling the country estates of English landed gentry, the eighteenth-century designs pioneered by Sir Humphrey Repton. Both Downing and Repton envisioned distant views of wooded groves that appeared natural but were actually edited and amended versions of nature. Such landscape gardens often included exotic trees, manmade lakes, and engineered geological features, all arranged with credible informality to resemble nature. Southern plantations were ideal sites for such gardens, providing both biodiversity and acreage for extensive cultivated woodlands.

Trees were cultivated near plantation houses, often emulating the landscape gardens of British landed gentry. The naturalistic groves combined native and non-native trees and included several fast-growing species introduced from Asia (*Harper's Magazine*, 1859).

Downing planned scaled-down landscape gardens for American homes, favoring the contrasting shapes and textures of "spirited" (pointed) evergreen conifers alongside deciduous hardwoods. Although the horticultural backbone of groves were native trees, Downing was intrigued by botanical introductions and suggested in his early writings that planters include exotic trees in their cultivated woodlands. Years later, in *Rural Essays* (1854) he favored native plants in cultivated landscapes, noting with regret that Americans often planted "cast-off nuisances of the gardens of Asia and Europe" rather than North American woody flora.[2] An essay in *The American Cotton Planter* offered similar landscape advice for plantations and villages, with particular encouragement to plant trees with evergreen foliage: "Of shade and ornamental trees, we have a number of suitable native sorts ... the ash, the oak, the wild poplar, the sycamore, the cherry.... Of ornamental trees and shrubbery, we still have a greater number of native origin. To my taste nothing of the kind looks so well, near a house, as evergreens. What can excel the pyramidal cedars, the verdant holly, the deep green large-leaved magnolia.... Who that walks around this town must but admire the vivid colors of these magnificent evergreens contrasting with the dull, dead and dreary appearance of the deciduous plants."[3]

Botanical biodiversity reflects Pleistocene glaciation, when encroaching ice sheets forced many plant species into southern regions of North America. The magnolias (*Magnolia* spp.) that characterize Southern landscapes were widespread across the northern hemisphere, and fossils (some as old as 100 million years old) have been found as far north as Greenland and Alaska. Other Southern landscape trees with northern roots include tulip poplar (also known as yellow poplar, *Liriodendron tulipifera*), silverbell (*Halesia carolina*), catalpa (*Catalpa bignonioides*), dogwood (*Cornus florida*), redbud (*Cercis canadensis*), and fringe-tree (*Chionanthus virginicus*). All of these were valued as shade trees with showy flowers, and they were transplanted and cultivated in plantation groves.

Groves of trees helped to mitigate rainfall during the spring and summer, controlling mud and improving road conditions within the plantation. Trees planted strategically in flood-prone areas made lawns and roads comparatively drier and more easily traversed. A large oak could uptake a hundred gallons of water daily, releasing it by means of transpiration (the loss of water vapor) into the atmosphere. In the extreme South, planters had particular preference for southern live oaks (*Quercus virginiana*), with a coastal range that extended from Virginia to Florida and along the Gulf of Mexico into Texas. Live oaks populated many plantation groves and although not consistently evergreen, they retained their leaves during cool months, dropping and replacing foliage in the spring. The trees transpired year-round, which may explain the frequent planting of live oaks along the avenues leading to plantation houses. Such shaded approaches were another English landscape tradition that was valued and replicated in plantation landscapes, beginning with the oak-lined avenues on South Carolina rice plantations. Both single and double ranks of trees were planted, depending on preference, carefully placed so that travelers viewed the homestead through an arched frame of branches.

Warm climate and high rainfall promoted rapid growth in live oak saplings, and tree-lined avenues became picturesque after a few decades. Their crooked branches were considered particularly evocative of the idealized southern landscape. In *A Journey in the Seaboard Slave States* Frederick Law Olmsted described such an avenue from a landscape perspective: "On either side, at fifty feet distance, were rows of old live oak trees, their branches and twigs slightly hung with a delicate fringe of gray moss, and their dark, shining, green foliage meeting ... densely overhead ... the arch was low and broad; the trunks high and gnarled, and there was a heavy groaning of strong, rough, knotty branches."[4]

Tree branches bore Spanish moss (*Tillandsia usneoides*), neither a true moss nor a parasite, but rather a flowering plant related to pineapples (Bromeliaceae, bromeliad family). Spanish moss thrives in humid habitats, where its hair-covered stems absorb water vapor and mineral nutrients without a root system. Although often associated with romantic images of plantation landscapes, Spanish moss had practical uses; fibrous clusters of the grayish stems trailed from live oaks and were easily collected. In Louisiana, plants were mixed with clay to make a traditional building material known as *bousillage*, and during the Civil War, encamped soldiers collected strands to fill mattress ticking. In *Resources of the Southern Fields and Forests* Francis Peyre Porcher recorded these uses and observations: "Great use is made in South Carolina of this plant when dried in stuffing chair cushion, mattresses.... It gives to the trees in winter quite a venerable and pleasing aspect, and is an indication of great moisture."[5] Spanish moss was also considered as a possible textile fiber, but it was not widely adopted. Gideon Lincecum, a southern naturalist, designed and built equipment for spinning and weaving Spanish moss into blankets to replace those taken by Union troops. The idea of using the plant as a fiber may have originated with the Native Americans who were his boyhood friends and companions.

In addition to live oak, water oak (*Quercus nigra*) and willow oak (*Q. phellos*) were favored for wet sites because the trees tolerated sodden, compacted soil. These deciduous species grew quickly regardless of soil type, but their cultivation in ornamental groves was a compromise because the trees dropped copious leaves and acorns. Planted together they hybridized widely with each other and with other red oak relatives, yielding hybrid swarms that have long vexed botanists and horticulturalists seeking to identify oak species. Ultimately many plantation oaks were cut for framing ships for the Confederate navy (see Chapter 8); live oaks were preferred for their durable and strong wood, but the timber required two or three years to dry thoroughly.

Andrew Jackson Downing praised beech (*Fagus grandifolia*) as a remarkable shade tree that "forms the darkest and densest shade of any of our deciduous forest trees,"[6] but some planters observed that cotton crops failed in fields that were once beech forests. Porcher cited research done by Justus von Liebig, suggesting that beech had a high requirement for calcium, a conclusion based on the high proportion of phosphate of lime (calcium pyrophosphate) in the ash from burned trees.[7] Beech may also have depleted soil potassium, based on the observation that the wood ash yielded high levels of potash (various potassium salts). Moreover recent studies suggest that beech may be allelopathic, releasing phytochemicals that diminish the growth of other species.[8]

Beech were widely planted in well-drained soils where the trees grew to four feet in diameter and provided a canopy of shade. Mature beeches met several wartime needs, and supplies were easily harvested from plantation landscapes. Confederate hospitals used beech leaves to fill mattresses, and the wood was a valuable fuel. Beechnuts were an emergency food for soldiers and livestock. The small, angled nuts are surrounded by spiny bracts that contain fagine; the mildly narcotic alkaloid is probably the chemical basis of the folk medicines made from beech which were used to treat headaches, fevers, tuberculosis, and melancholia.

Other native trees that were intentionally planted and cultivated on plantations included eastern red cedar (*Juniperus virginiana*), American holly (*Ilex opaca*), and various pines (*Pinus* spp.). There was confusion in the vernacular names of junipers and cedars; although the genus *Juniperus* has aromatic wood, it is distinct from the genus *Cedrus*, the true cedars that are native to the Old World. Some groves and lawns were planted with the non-native cedar of Lebanon (*C. libani*), which was cultivated in England by the 1660s and coveted by American gardeners for planting alongside native red cedars and other evergreens. Red cedars had the "spirited" shape admired by Downing, and the trees could be transplanted from wild areas with poor soils. The first edition of *The American Cotton Planter* described *Juniperus* as "a beautiful conical bush, grouping and combining well with Cedars and other species" and noted that holly was suited "for lawns and small groups, on account of its shining green leaves and numerous spikes of coral berries."[9] Eventually some juniper groves were felled for making wartime huts and shelters; the aromatic branches were used as bedding in military hospitals, where decoctions of the berries and needles were used medicinally for rheumatism and blisters.

Also known as red cedar, juniper trees were transplanted from the wild for landscaping southern homesteads. During the war, medicinal decoctions were made from the needles and berry-like cones, and the aromatic branches were used in hospital beds (author's collection).

Southern magnolias (*Magnolia grandiflora*) were the most iconic ornamental trees in the plantation landscape, with a range extending from North Carolina south into central Florida and Texas. Wild trees grow near water, including swamplands and coastal dunes, resulting in the folk wisdom that magnolias might be an effective cure for malaria. However, standing water and extreme heat were not essential for their culture. Southern magnolias were successfully cultivated in groves and gardens across the South (and later as far north as New England). Saplings grew quickly in the Southern climate, achieving heights of a hundred feet and diameters of two feet.

Plantation magnolias were admired

as specimen trees with handsome foliage, and the scented flowers symbolized antebellum traditions. The spring exhibition of the Mobile Agricultural, Horticultural, and Floricultural Society included a "horn of plenty ... composed of evergreen wreaths interlarded with rare and beautiful flowers, the apex a magnificent bloom of the magnolia grandiflora."[10] Some planters chose magnolias for the shaded avenues leading to plantation houses, where some later anecdotal notes recorded that Confederate soldiers used the trees to construct corduroy roads.

The British naturalist Mark Catesby introduced southern magnolias to England in the early 1700s, where they were cultivated at the Chelsea Physic Garden for possible medicinal qualities. Tulip poplar trees (*Liriodendron tulipifera*) are also members of the magnolia family (Magnoliaceae), another native species that colonized swamps in nature but thrived in plantation groves and lawns. In forested habitats, the trees reach 160 feet in height and several feet in diameter; like magnolia they were widely regarded as a treatment for rheumatic pains and substitute for imported feverbark (*Cinchona* spp.), the source of quinine used to treat malaria. Tulip poplar trees were introduced as curiosities to England in the 1660s, one of the first botanical imports from the American colonies.

Asian Imports

By the early nineteenth century, American nurseries advertised many Asian trees for cultivation, often with appealing traits such as rapid growth and showy flowers, although rarity and curiosity were factors as well. Then, as now, horticulturalists were intrigued by relatively unknown plants, and nurserymen moved quickly to propagate and introduce recent Asian discoveries into plantation landscapes. The planter class desired landscapes that reflected wealth and status, and in the Southern climate Asian trees produced fast results. However, many imports grew aggressively by self-seeding or asexual suckers or rhizomes, and they soon naturalized and permanently invaded southern habitats.

An early source for many Asian introductions was the nursery established in 1786 near Charleston, South Carolina, by the French botanist Andre Michaux, who propagated and marketed several ornamental species cultivated at the Jardin des Plantes in Paris. Michaux introduced to antebellum landscapes iconic "southern" trees including camellia (*Camellia japonica*), crape myrtle (*Lagerstroemia indica*), mimosa (*Albizia julibrissin*), and Chinese parasol tree (*Firmiana simplex*), which all thrived in warm American climates at latitudes close to their native habitats. All were small Asian forest trees that were grown as specimen trees in plantation gardens. The genus *Camellia* includes cultivated tea trees (*C. sinensis*), but *Camellia japonica* was prized for its abundant flowers. The trees are native to China, Japan, and Korea and have showy petals in shades of white, pink, and red. Occasional mutations resulted in double flowers which were propagated as new cultivars. Probably because of the flower shape and tendency to double, some likened camellias to roses, but they are not closely related. Crape myrtle was native to China, despite the epithet *indica* in its scientific name. Clusters of small flowers were white, pink, red, or purple, depending on the variety, and they appeared even in the heat of summer. The trees tolerated high temperatures and drought conditions, and they were easily propagated from suckers.

Mimosa trees had delicate pink flowers that appealed to many plantsmen, including William Bartram, who sent seeds of "a beautiful flowering tree ... lately brought to us by the celebrated Michaux the elder" to Thomas Jefferson.[11] Its flowers have showy stamens rather than pigmented petals; a reference to their silky appearance, the epithet *julibrissin* is derived from a Persian word that translates as "silk flower." The trees grew quickly, flowered throughout the summer, and cast dappled shade through their double-compound leaves. Typical of other legumes, the roots of mimosa are hosts to endosymbiotic bacteria (*Rhizobium* spp.) that can convert atmospheric nitrogen into usable compounds, allowing the trees to live on nutrient-poor soils such as old fields and abandoned land. Mimosa seeds have remarkable longevity and often germinate in disturbed sites, which along with rapid growth and high seed production are traits typical of invasive species. Once a prized tree in plantation landscapes, mimosa now self-seeds and colonizes aggressively across the South and as far north as New York and Connecticut.

A similar situation occurred with royal paulownia (*Paulownia tomentosa*), fast-growing trees that were favored for their upright clusters of tubular purple flowers. Early nurserymen compared the species to catalpa trees (they are related although now placed in separate families), and William R. Prince's *1844 Descriptive Catalogue* included paulownia as "a new ornamental tree of great beauty, splendid foliage ... a profusion of beautiful purplish bell-shaped flowers." However, plantation growers soon realized that seedlings grew anywhere, including disturbed sites and poor soils. A metabolic adaptation explains how the trees survive in extreme heat and drought; paulownia photosynthesizes using the C4 pathway usually found in tropical monocots such as sugar cane and pineapples (see Chapter 3) in which carbon dioxide is absorbed and stored as part of a four-carbon compound. C4 metabolism works efficiently at high temperatures when a process known as photorespiration can interfere with efficiency, explaining how paulownia trees thrive in extreme heat. Seeds also spread by an unexpected means; imported Chinese porcelain was packaged and shipped in layers of the soft winged seeds, which scattered along train tracks if the crates broke apart. By the end of the nineteenth century, the trees had naturalized widely and lost their appeal to planters. Paulownia trees vanished from cultivated southern landscapes as they became invasive on abandoned land and disturbed sites across the South.

Chinese parasol trees (*Firmiana simplex*) have large leaves that resemble those of maple or sycamore, but the species is a member of the mallow family (Malvaceae) and a relative of cotton and hibiscus. The shiny foliage forms an umbrella-like canopy that was valued on plantations, but its small flowers self-pollinated and produced thousands of seeds that spread widely, germinating into seedlings that grew invasively. The fast-growing saplings developed deep taproots that prevented the trees from uprooting, which had been a problem with the mature chinaberry trees prized on many plantations. Chinaberry (*Melia azedarach*) was also planted extensively and was one of the woody species that typified southern landscapes; its compound foliage shaded many of the avenues leading to plantation homesteads. In *Uncle Tom's Cabin*, Harriet Beecher Stowe described the trees at Simon Legree's plantation, "a noble avenue of China trees, whose graceful forms and ever-springing foliage seemed to be the only things that neglect could not daunt or alter." In "Evangeline," Longfellow mentioned planters' houses shaded by chinaberry trees. Once known as the pride of India, the trees are small, fast-growing relatives of tropical mahogany trees (family Meliaceae). Chinaberry produces spring clusters of violet flowers, followed by yellow berries which are eaten by birds. Planters observed that the trees resisted insect attacks, and some suggested using "china leaves and berries" to repel insects from fruit trees.[12]

William R. Prince's *1790 Broadside of Fruit Trees and Shrubs* was the first nursery to offer chinaberry for sale to American growers, and they soon became as popular as lilacs were in the North. By 1857, the *Southern Cultivator* described, "This beautiful shade tree, under whose wide-spreading branches the Southern People spend so much of their leisure time in the hot summer ... one of the greatest blessings of Providence."[13] Olmsted mentioned "pride-of-China" as a shade tree planted among cottages and cabins.[14] However, gardeners and horticulturalists became frustrated with the continual rain of leaves, twigs, and berries. Its popularity as a cultivated plant dwindled as wild populations naturalized across the South, although the seeds had a practical use as a source of oil for soap-making during the war years (see Chapter 5). As with many other species that are now considered invasive, chinaberry seeds germinate easily on poor soils, develop shoots from their roots, and form dense drought-tolerant thickets in uncultivated areas or abandoned fields.

Chinaberry trees lined the avenues on many plantations where they were valued for providing dense shade in a hot climate, but the trees quickly naturalized on poor soils. The seeds provided fats for soap-making during the war years (morphart © 123rf.com).

Orange trees (*Citrus sinensis*) were planted in groves and along avenues on plantations in the extreme South. The low-branching trees had evergreen foliage, fragrant white flowers, and edible fruits known commonly as sweet oranges. They evolved into an image of the South even for those who had never visited the Confederate states; in *Uncle Tom's Cabin*, Harriet Beecher Stowe described a semi-tropical New Orleans garden in which "two large orange-trees, now fragrant with blossoms, threw a delicious shade" and the walks that Uncle Tom and Eva took "under the orange-trees in the garden." Olmsted took particular interest in oranges cultivated outdoors, and he noted a South Carolina plantation house surrounded by a grove of fruit-bearing trees and successful orange crops raised in Alabama "grown with some care, and slight winter protection."[15] Orange trees are entirely different from Osage orange (*Maclura pomifera*), a native shrub used in planting hedges.

Tree of heaven (*Ailanthus altissima*) is perhaps the most notorious of the invasive plants resulting from Asian introductions into southern landscapes. In 1751 a Jesuit priest sent its seeds from Peking to the Chelsea Physic Garden, and by the 1780s the trees were cultivated in Philadelphia gardens. Prince's *Catalogue of Fruit and Ornamental Trees* (1832) offered the trees as specimens with "quick growth and handsome foliage." Their popularity caught on initially; Olmsted described the plantation house of a Virginia tobacco grower as "an old family mansion, which he himself remodeled in a Grecian style.... A number of the old oaks still stood ... and there had been some in its front.... These, however, he had cut away as interfering with the symmetry of his grounds and in place of them had planted ailanthus trees in parallel rows."[16] Tree of heaven was widely planted along avenues and in groves and ornamental plantings, but plantation dwellers were soon annoyed by the scent emitted by all parts of the tree. Saplings grew at an alarming rate (often over six feet per year), and once established, a tree could clone itself aggressively with dozens of root suckers. By the end of the nineteenth century, the once-prized tree of heaven was widely regarded as a noxious woody weed.

Orange trees had evergreen foliage, fragrant white flowers, and edible fruits known commonly as sweet oranges. With some protection, crops grew on plantations in the Deep South, and the trees were valued for providing garden shade (Harter, 1998).

The most botanically unique Asian tree in cultivation was *Ginkgo biloba*, also known as the maidenhair tree, named for the resemblance of its leaves to maidenhair ferns (*Adiantum* spp.). Botanically speaking, ginkgo is a gymnosperm, seed-producing trees that are more closely related to conifers than flowering plants. They are unrelated to other living trees and thus are classified as the sole species in a genus (*Ginkgo*), family (Ginkgoaceae), order (Ginkgoales), and division (Ginkgophyta). Fossils date the evolutionary line to the Permian, and the extant trees are now regarded as "living fossils." The trees were revered in China since ancient times, where they were cultivated in temple gardens and valued for their medicinal properties and edible seeds. While related fossils have been found across the Northern Hemisphere, the only living ginkgoes originate from China. However, probably none of this was known to nineteenth-century growers who cultivated maidenhair trees as arboreal specimens on southern plantations; seedlings grew slowly into stately forms with golden foliage in the fall. Those who read Andrew Jackson Downing knew that he regarded ginkgo as "a great botanical curiosity ... so singularly beautiful with its fern-like foliage, that it is strikingly adapted to add ornament and interest to the pleasure ground ... we would recommend that it be planted near the house, where its unique character can be readily seen and appreciated."[17]

However, planters soon realized that ginkgo is dioecious (pollen and seed production occurs on separate trees), and whether a particular tree produces pollen or seeds only becomes known when a tree is fairly well established. The seeds have a distinctive odor that results from the concentration of butyric acid produced in the outer fleshy layer of the seed coat. In warm climates, the rancid scent of mature ginkgo seeds was notoriously unpleasant, so nurseries learned to propagate pollen-producing plants from cuttings to avoid malodorous seeds in gardens and cultivated groves.

Hedges

Antebellum plantations varied in size from a few hundred acres to vast parcels of over a thousand acres, and agriculture on this scale required fences to define uses and demarcate land boundaries. Early planters used rail fences to protect gardens, groves, orchards, work yards, pastures, and fields, but widespread clearing of land resulted in dwindling timber stock. Thus some planters began to develop hedges as natural barriers. Inspiration may have come from the hedgerows that separate farm fields in England, in the same way that antebellum plantation landscapes reflected other aspects of British tradition, design, and ownership.

Plants grown as hedges had to be drought resistant and anatomically tough or spiny, and several such native shrubs and small trees were known to survive in the varying conditions of southern fields. The cassine holly (*Ilex cassine*) has a natural range from coastal Virginia south to Florida; the robust, evergreen shrubs have tough leaves with occasional spines and red berries, making the plants ideal for ornamental and agricultural hedges. In *Hedges and Evergreens* (1859) John Warder described cassine as "a low, evergreen tree ... the berries are large in proportion, red and persistent during the Winter—as they are not eaten by birds, they produce a pretty effect."[18] He also recommended the more widespread North American *Ilex opaca* as "a beautiful evergreen hedge, and a pretty good barrier against cattle" and English ivy (*I. aquifolium*) for its "armed green leaves and bright-red berries."[19] In southern climates, hollies matured into small trees, but closely planted saplings could be clipped into impenetrable hedges several feet in height. Southeastern yaupon hollies (*I. vomitoria*) were also widely planted as evergreen hedges, but the shrubs lack spiny leaves, and deer sometimes browse on the twigs and foliage. Long before plantations were cleared from southern forests, yaupon was familiar to Native Americans who infused the leaves into a tea containing caffeine and theobromine.

In particular, resilient hedges were needed to surround agricultural fields. Some were planted with a Eurasian pyracantha (*Pycracantha coccinea*), thorny shrubs that tolerated dense clay soil and closely resembled the hawthorns used in planting English hedgerows. The species was first classified as an ornamental hawthorn (*Crataegus* spp.), which was valued for its red berries and ability to grow against walls, and only later used for plantation hedges. Pyracantha was adopted after the failure of many hedges planted with English hawthorn (*Crataegus oxyacantha*), a problem more likely related to poor management and pruning rather than the botanical nature of the shrubs.[20] The native red cedar (*Juniperus virginiana*) tolerated a wider range of various soils, moisture conditions, and temperature extremes across the South. Left untrimmed, red cedar grew into trees, but they could be pruned to produce dense hedges. The leaves on young plants are sharp-pointed needles that are replaced by scale-like foliage as red cedars mature; perhaps it was this trait that turned attention to shrubs with permanent thorns.

Planters ultimately preferred Osage orange (*Maclura pomifera*) and Cherokee rose (*Rosa laevigata*, native to China despite its vernacular name) for hedge-making. The latter bore attractive white flowers and was easily propagated from seeds, roots, and cuttings, a quality important for mass planting. Olmsted praised its glossy, evergreen foliage and described Cherokee rose as the most valuable shrub for establishing hedges.[21] Cherokee roses were often planted along rail fences, which provided support until the plants formed a sprawling natural hedge. Grapevines, green briar, and other wild plants that self-seeded and became established among the thorny roses.[22]

Osage orange offered all of the most admirable traits of a hedge plant. First introduced to the eastern United States by Meriwether Lewis and William Clark, the species was native to the central states and easily propagated. The small trees bear robust thorns and large fruit that superficially resemble true oranges. Periodically pruned to shrub height, the trees formed a dense hedge that resisted animals and insect pests, a trait attributable to the presence of the naturally-occurring repellent 2,3,4,5—tetrahydroxystilbene. Agricultural journals advertised Osage orange to Southern planters, who in the 1850s may have been unfamiliar with the species. A note in the *American Cotton Planter* offered the services of an Ohio nursery in planting miles of "the most desirable of all live fences" in Alabama.[23] In an effort to educate planters, the 1855 Alabama Agricultural Society fair offered an award for hedging plants, most notably Osage orange "with a description of planting, trimming and training the hedge."[24] Experienced growers advocated growing Osage orange from seed and then transplanting as needed; growing advice in *The American Cotton Planter and Soil*

Osage orange was introduced from the central U.S. by Lewis and Clark. Southern growers adopted the small trees as insect-resistant hedge plants for plantation fields; hedgerows were planted from seed and required early pruning and management to create a dense barrier of thorny vegetation (author's collection).

of the South provided detailed instructions: "Procure good seed—soak them until slightly swollen, then sow in a nursery as you would apple seed ... remove them from the nursery to the hedge-row ... six to eight inches apart in the drill ... your hedge may be turned loose to take care of itself at three years old."[25] However, care and management were essential; in response to heavy pruning, the roots grew suckers that needed to be interwoven among the branches to close gaps and create a dense hedge.

Antebellum agriculturalists debated the comparative farming and financial merits of Cherokee rose and Osage orange in replacing rail fences. Farm journals published firsthand accounts of hedge successes and failures and detailed advice based on practice and experience. Some growers maintained that the roses grew better in shaded conditions, but they were more difficult to prune. In contrast, Osage orange may have depleted the soil of nutrients, but the young plants were more easily trimmed and coaxed into dense hedges, although neglect could result in hog-sized holes among the branches. Osage orange shrubs predated and presaged barbed wire, which was not patented and widely used until the late 1860s. In fact, the structure and sharp embellishments of barbed wire functioned much like the formidable thorny suckers of Osage orange.

Botanists, however, focused on the unique fruit of the Osage orange, which were often known in the South as hedge apples. Typical of the mulberry family (Moraceae), each cluster of flowers matures and fuses into a single multiple fruit with a textured surface and bright, yellow-green pigmentation. These have a tough texture and massive size (about the size and shape of a modern softball), and most birds find them unpalatable. In the antebellum South, most Osage oranges fell to the ground, where a few may have been eaten by horses. Another dispersal mechanism must explain their wide range across the plains and prairies of the central United States. Native Americans valued the wood of Osage orange for making hunting bows, and the wood and shrubs may have been widely traded from its original range. Meriwether Lewis sent cuttings to Thomas Jefferson, with a note (March 26, 1804) that explained, "So much do the savages esteem the wood of this tree for the purpose of making their bows, that they travel many hundreds of miles in quest of it."

Shrubs of the thorny Cherokee rose were native to China and known for their attractive foliage and white flowers. New plants were easily propagated from seeds, roots, and cuttings to create sprawling hedges around agricultural fields (author's collection).

However, pollen records from glacier-free times of the Tertiary period reveal an even wider range across North America for ancestral Osage oranges. Daniel Janzen, an evolutionary botanist, has suggested that the only animals large enough to consume the fruit were woolly mammoths, mastodons, and ground sloths, the extinct megafauna that disappeared by the end of the Pleistocene epoch about 13,000 years ago. Referring to such fruit as anachronisms, Janzen has postulated that Osage oranges adapted through natural selection to the throat dimension of megafauna that roamed North America in the relatively recent past; they are the only known animals that could have swallowed the tough-textured whole fruit.[26]

Brush Arbors

Planters encouraged enslaved people to practice Christianity, and simple wooden churches were part of the plantation landscape. Enslaved people also held on to traditions that they recalled from Africa, which were often preserved and practiced in sheltered areas known as brush arbors, also known as hush harbors. These were secretive, secluded sites to sing and pray after dark, as a way to pass folk wisdom and practices on to those without memories of African tradition. The services often combined Christian hymn-singing, evangelism, and sermonizing with African songs and stories, a seemingly innocuous agenda. Personal reminiscences include references to brush arbors, including the recollections of Robert Anderson, whose enslaved father used the New Testament to preach and teach in his brush arbor.[27] Enslaved people also used the time to recall and exchange herbal knowledge and medical treatments, in particular practices that reflected African rather than North American botany.

Some plantation owners willingly allowed brush arbor meetings, while others feared that enslaved people might share grievances and plan revolts. Indeed those caught at meetings risked beatings and other punishments. Gatherings were held in a purpose-built arbor at an established site, usually a nearby hollow or forested tract of land. The arbor resembled a lean-to constructed from poles and branches (brush), with logs or rough boards as seats.[28] The roof was assembled from branches, sometimes insulated with layers of hay, and they resembled the arbors (or harbors) built by white Christians as shelter during lengthy camp and revival meetings.

Trees used in constructing brush arbors were native to the region and landscape, often oaks or pines or whatever other species were common. Meetings were potentially noisy events with singing, preaching, and shouting, but foliage provided both shelter and sound-deadening effects. Layers of brush, depending on their thickness and the number of leaves, may have lessened noise by half; during winter, evergreens deadened sound more effectively than deciduous branches. Plant material camouflaged the sites and silenced voices, keeping the meetings private and oral traditions intact between generations.

Wartime Landscapes

The agricultural wealth of the South was concentrated in plantations, complex landscapes that consisted of domestic, ornamental, agricultural, and wild areas. Land use was flexible; fields were abandoned when soil nutrients dwindled, and new acreage was frequently cleared. Untended woodlands and groves grew densely as trees reseeded, and other uncultivated sites included swamps and other wetlands. Such wild areas on plantation land became hiding areas for valuables including jewels, silver, and food and cotton crops that owners wanted to keep from the hands of Union troops. Even birds' nests were co-opted to hide small items that civilians wanted to protect from foraging soldiers.

Union troops habitually raided southern plantations during the latter years of the war. As a military strategy, the destruction of plantations undermined agriculture, resulting in food shortages and hunger. The type of foraging raid ordered by Union officers was known as a *chevauchée*, a remnant of French medieval warfare planned intentionally to ruin crops, destroy landscapes, and generally cause havoc for civilians. By attacking plantation landscape and economy in *chevauchée* raids, Union troops undermined and diminished the Confederacy—both symbolically and practically. Union troops pillaged widely as a planned military strategy. Soldiers commanded by Grant and Sherman destroyed corn fields and stole provisions to supply the Union army; their intention was to carry minimal rations and to forage for necessities as the campaigns progressed.

Plantation landscapes were summarily destroyed as fields were cleared of crops and trampled and fences and houses burned, despite the concern even among Union generals that the strategy encouraged barbarism among troops.[29] Plant communities were often collateral damage during the foraging raids, although hedges sometimes played a defensive roll; at least two Confederate defeats resulted from Osage oranges in the landscape. On June 7, 1863, troops commanded by Major General J.G. Walker attacked Union troops at Milliken's Bend, Louisiana, but they were stopped in place by an impenetrable Osage orange hedge surrounding the village. Union soldiers shot through the dense growth while sheltering safely behind the hedge. Positioned near Franklin, Tennessee, in November 1864, Union troops again adapted Osage orange shrubs for defense. Under the command of General John M. Schofield, soldiers excavated trenches behind dense growth of Osage orange, most likely abandoned hedges that naturalized into thickets. Armed with repeating rifles (which replaced muskets), they fired with ease at the Confederate soldiers from the 35th Alabama Infantry Regiment who tried to advance across the open fields.

3

Agriculture and Crops

North and South

At the start of the Civil War, more than half of all Americans lived on farms. Farming was a way of life in both the North and the South, despite differences in land, soils, crops, and labor, but cash crops on a massive scale were strictly southern. Agriculture in antebellum Confederate states ranged from small subsistence farms to high-acreage plantations of sugar, rice, and cotton, concentrated in coastal areas, river valleys, and delta regions of the lower South. Three-fourths of all American exports were cash crops, largely products cultivated on plantations. In contrast, family farms in northern states were typically moderate in size, cultivating a mix of cash crops and household needs; yet these households also depended on cotton and sugar grown and harvested using enslaved labor. In short, antebellum agriculture was an economic weave of regional interdependence. The market for plantation crops was widespread, despite contempt for slavery and Abolitionist promotions of alternatives such as linen and maple sugar to replace slavery-dependent cotton and sugar.

The Civil War shifted Southern agriculture away from the monoculture of cash crops and toward subsistence farming on land that represented more than acreage. Indeed agriculture defined the social order of the South. Speaking in 1852 to the Black Oak Agricultural Society of South Carolina, botanist Henry William Ravenel had observed, "An agricultural people are always more strongly attached to the soil on which they have been reared. They become identified with it." A decade later, arable land was essential for autarky as food shortages burgeoned and the Confederate government encouraged planters to grow staple foods rather than cash crops.

Once Union forces blockaded Confederate ports, cotton could not be shipped to northern mills, and the excess bales accumulated in warehouses and ports. Southern newspapers published the motto "Plant Corn and Be Free, or Plant Cotton and Be Whipped." Faced with civilian and military demands, the strategic option was to convert from a plantation economy to food farming and preservation because corn cultivation meant bread on southern tables. Thus during the first year of the war, the cotton harvest in Confederate states dropped from 4.5 million bales (1861) to 1.6 million (1862); sugar production plummeted from 459 million pounds to 87 million pounds during the same years.[1] Most planters willingly reversed the antebellum ratio of three acres of cash crops for each acre planted in corn; typical wartime farms had 600 acres of cornfields and 200 acres of cotton, the latter planted as a non-perishable commodity that could be stored and sold once the blockade ceased. Some planters nevertheless continued to cultivate vast cotton crops despite Confederate instructions to grow corn and other food crops. Risking charges of treason, they were willing to profit by sales to speculators and buyers from the North or smuggle their crops to Europe by way of Texas and Mexico.

Ironically, generations of enslaved people had traditionally planted gardens of vegetables and herbs to provide for their own families, and planters had to do the same to survive in wartime. Enslaved workers cut trees and cleared land, tilled soil, collected and spread manure, cut fodder, and cultivated and harvested crops. Their knowledge of applied agriculture was extensive, acquired by interacting with plants and soils, observing growth, and sowing seed for the next season. Enslaved labor had fueled antebellum agriculture, and without this source of free energy and practical knowledge, the system ultimately crumbled.

During the antebellum years, Southern planters organized and joined agricultural societies, organizations that promoted agrarian ideals and shared information and improvement schemes through farm journals, including *The Cotton Grower*, *Southern Agriculturalist*, *Farmer and Planter*, and myriad others. Such organizations became popular in the 1830s, and in South Carolina, for instance, they tripled in number between 1826 and 1847.[2] Agrarian groups organized festivals, sponsored lectures, sponsored lavish dinners, laid on sumptuous picnics, and encouraged friendly competition. Their major role was to promote unity and social standing among the planter class, even as plantation agriculture became unsustainable for lack of free labor.

The pro-slavery stance was embedded in southern agricultural journals, amidst articles on soil structure, cotton varieties, and useful farm machinery. Publications disseminated practical scientific knowledge and farming technology, promoted the awareness of social class, and pitched a *status quo* message to landowners whose way of life depended on plantation slavery. Independent publishers produced farm journals such as the *Southern Cultivator*, which covered diverse topics in culture, literature, politics, and agriculture aimed at planters and their families; the January 1861 issue included articles on home hygiene, dog breeds, school exercise programs, bread baking, soil enrichment, cotton culture, horticultural hints, and poetry. Subscriptions were cheap at about a dollar a year. Although some forward-thinking articles were regarded with skepticism, planters relied on advice from the *Southern Cultivator* in preparing for war. *DeBow's Review* presented a similar pro-southern, secessionist perspective and included articles on commerce, economy, agriculture, and the southern way of life.

Many southern growers ignored farm journals and carried on with practices that damaged soil. Advice on crop rotation and deep plowing went unheeded, and as a result, soil nutrients diminished with each crop, and tons of topsoil washed away with seasonal rains.[3] In his diary of 1843, Edmund Ruffin, the state agricultural commissioner of South Carolina, referred to the "barbarous usage" of soil planted with cotton year after year, so there was clearly awareness of nutrient-depleting crops.[4]

Planters sometimes alternated one cash crop with another, a practice that did little to restore soil nutrients, but there was still widespread optimism for wartime farming despite diminishing soil fertility. As late as 1863, Porcher confidently predicted, "That system of husbandry which accumulates the elements of crops and fertility in every acre cultivated, is still a myth to most planters. Southern nationality will expose, and happily correct many errors. We shall learn to make as much cotton and corn on two acres as we now do on six, and at the same time we shall produce tenfold more of the necessaries and comforts of civilized life."[5]

Plantation practices stood in sharp contrast to crop rotation in the small farms that cultivated family food rather than cash crops. *The Country Gentleman*, published in upstate New York and described as a "journal for the farm, the garden, and the fireside," recommended a four-year crop sequence for arable fields beginning with corn, potatoes, and carrots, followed by barley, peas, and beans; wheat and corn; and lastly clover.[6] Three of the recommended crops on alternating years (peas, beans, and clover) are legumes. Although the first *Rhizobium* bacteria were not isolated until 1889, there was practical knowledge of the beneficial effects on soil fertility of planting legumes with symbiotic nitrogen-fixing microbes.[7] An agricultural correspondent to *The American Cotton Planter* recommended a rotation of cotton, grains, and peas, with the pea vines being plowed into the soil at the end of the growing season.[8] Yet although such wisdom was known, it was often ignored in the South. Agricultural journals referred to pea vines as the "clover of the South" (referring to the ability of plowed-in clover to improve soil fertility), but few followed this sort of advice until the Civil War resulted in food shortages. With war on the horizon, detailed instructions on corn cultivation included advice on planting pea vines among corn stalks as a source of soil nitrogen.[9]

Correspondents to agricultural journals debated the relative merits of applying manure and other soil amendments on the surface or plowing them deep into the soil. Buried manure fermented, and the heat produced by soil microbes killed viable seeds so that weeds were not spread, yet there were circuitous arguments to support both practices.[10] Some planters opposed the use of plows, but shallow cultivation with hoes alone prevented nutrients from penetrating deep soil layers, and crops suffered. It was avoidance rather than disregard that caused these problems. In the South, one anonymous writer noted, "If there had been ... proper rotation of crops ... if manure of the barn-yard and the wood-ashes of the house fires had been husbanded ... those fields now looking so barren and forlorn would have been more fertile."[11] The report of the federal Commissioner of Agriculture of the newly created U.S. Department of Agriculture related shallow plowing and nutrient depletion to slavery, noting, "It is remarkable that this result generally follows in all slaveholding countries, owing doubtless to the negligent and unthrifty habits which always attend that unrighteous practice."[12]

The results of poor soil management were familiar, and some even imagined that soil affected heredity. An article the *American Cotton Planter* noted, "When cultivated plants are neglected and allowed to grow in a poor soil, they soon revert to their wild condition. It therefore requires a continuance of suitable conditions to perpetuate those peculiarities that render them useful to man."[13] Indeed when cultivated varieties exhibited scraggly growth in low-nutrient soils, it may have seemed that growth conditions caused plants to revert to their wild types, but inheritance was not yet understood scientifically. Between 1856 and 1863, the Austrian monk Gregor Mendel investigated heredity by hybridizing garden peas; shortly after the Civil War (1866) he published a paper on his work, but its significance in understanding genetics was not widely understood until the early twentieth century.

Southern planters thought of land as a limitless resource. As Frederick Law Olmsted observed during his southern travels, only about one-fourth of available land was farmed in the South, compared to nearly half in the North. When productivity declined, southern farmers moved on to untilled fertile acreage, while their Northern counterparts practiced progressive methods of soil management. The 1862 *Report of the Commissioner of Agriculture* described Massachusetts as a state "unfavored in soil and climate" but remarkable because "knowledge is power ... to the end that labor may be more productive because more skilled and better directed.... The farmer should have taste to appreciate and enjoy the beautiful in nature and in art; taste to adorn his home and his lawns with shrubbery, flowers, and works of art; taste to admire the ripening fruits, the glowing landscape, the processes of nature."[14] By the 1860s, there were proposals to include agriculture in the curriculum of northern common schools, in addition to basic literacy and arithmetic. According to its proponents, practical instruction using *Nash's Progressive Farmer* and similar texts in the "systematic management of the farm, with economy in manure and its cultivation" was essential for success on the land.[15]

Problems mounted as the war progressed, with intense agricultural and economic impact in the South. As a result of cash shortages, the Confederacy offered bonds that could be bought with crops rather than cash. As a result, the government accumulated stores of cotton and tobacco, but it became difficult to sell these to Europe for cash to pay for war goods. In addition, the Confederate government levied a 10 percent in-kind tax on corn, wheat, oats, rice, sugar, sorghum, potatoes, beans, buckwheat, peas, cotton, tobacco, fodder, and hay. Taxation was a hardship for subsistence farmers, who also lost crops as troops on the march pillaged fields and crops. In mid–1863, the Confederate Congress passed an in-kind tithe tax to provide provisions and forage for troops, allowing the confiscation (impressment) of 10 percent of the grains, fodder, potatoes, legumes, sugar, and tobacco grown for market. Impressment officers seized food and cash crops at levels that often exceeded 10 percent, resulting in more shortages and financial hardships for civilians. Officers confiscated essential tools, including shovels, saws, and pick-axes, and many impressed goods rotted because Army depots lacked suitable shelter. In contrast, Union army quartermasters purchased crops from Northern farmers, who benefited financially.

The 1860s were a time of mechanical innovation in agriculture, and a wide range of labor-saving devices freed men to serve in the military. Historians have debated the impact of war on the development of farm machinery; did the war stimulate invention, or did inventions enable military service? The pathway to agricultural inventions was already in progress, so the latter seems more likely. Women left alone to manage family farms were desperate for help with seasonal work, and agricultural machines were the best option for those who could afford and maintain them. These inventions pre-dated the war, but labor shortages stimulated their widespread adoption when the increased price of wartime agricultural products resulted in ready cash.

Time was of the essence in bringing in crops because mature grain lasted about a week before the heads shattered or rain ruined the entire field. The process involved cutting followed by threshing, the process in which the seeds were separated from the outer grain layers and stems. Some machinery did both jobs; horse-drawn combine harvesters were developed in Scotland in 1826, and by the Civil War, they were used in American fields to reap and thresh grain. Hay was needed for farm animals, and its collection was simplified by sickle-mowers with toothed blades that alleviated hand-scything. Horse-drawn hay rakes, also known as sulky rakes, were advertised for use by women. However, wartime shortages of metal and labor made it difficult or impossible to replace worn or broken parts of harvesting machinery, plows, hand tools, and other implements. Scythes, shovels, axes, and hoes became increasingly scarce as southern manufacturers converted their factories to making essential materiel rather than farm machinery. This had a drastic effect on poor farmers in particular, who often had few resources and no duplicate tools on hand; cast-off tools and machinery were brought from barns and put back into use. Northern farmers were able to purchase farm machinery for the duration of the war from manufacturers in New York, Ohio, and other Union states, and their harvests reflected the benefits of mechanized agriculture.

A further complication was the dwindling supply of horses, the result of poor husbandry, fodder shortages, and military demand for horses and forage. The Confederate and Union armies lost hundreds of horses each day, which needed replacement, and each horse required a daily supply of hay and grain. Farm machinery required the strength and stamina of work horses to traverse agricultural fields. Without machines and a way to move them, southern farmers often turned to relief handouts as a means to feed their families. By the end of the war in Confederate states, need crossed class lines when the families of planters also encountered shortages and hunger. In deep contrast, Northern farmers benefited from agricultural improvements, increased productivity, and wartime inflation; their farms supported families and soldiers and also provided grains and other commodities for export.

Soils, Nutrients, and Cultivation

Southern agriculture began with the clearing of ancient conifer or hardwood forests. Crop cultivation followed, but soil nutrients were sometimes limited—despite the impressive primary forests that once blanketed the agricultural fields. Soils were often the red clay that forms when minerals weather in a humid temperate landscape, typical of the southern soils now classified as ultisols. They are known for their high soil acidity (pH often less than 5) and comparatively low levels of essential nutrients such as calcium and potassium. Agricultural problems stemmed from the soil; ultisols resist continual cultivation and require fallow periods for soil fertility to return. Some growers learned that burning trees and woody undergrowth helped to replace mineral salts, and ash was used as a common soil amendment to replace the calcium depleted from southern ultisols.

Nevertheless soils were often depleted after five years of cultivation. The effect of intensive monoculture on southern soils was extreme; early, mid-season, and late plantings damaged soil structure and diminished the mineral salts that provided plant nutrients. Old fields were often abandoned for decades and sometimes later reclaimed for cultivation. Given the size of plantations and prosperous farms, land seemed like a limitless resource; there was seemingly always more acreage to clear, plant, and cultivate. However, land-clearing required hard labor to expose virgin soil. Enslaved people cut and burned forest trees, dug out their stumps and roots, plowed furrows, built fencing, and carried in the manure used to fertilize crops. Land-clearing was continuous, as was the practice of abandoning old fields to grazing cattle and self-seeding wild plants. Indeed abandoned land typified all economic levels of Southern agriculture, even when fields were cultivated in subsistence crops. Appalachian farmers left charred trunks and roots in place to control erosion in sloping

Intensive farming soon depleted southern soils of nutrients, and grazing animals occupied abandoned fields. New acreage was frequently cleared for cultivation in virgin soils; enslaved workers did the arduous tasks of clear-cutting, stump removal, fence-building, and plowing (Library of Congress).

corn fields, but they also continued to clear land when soil nutrients diminished and crops dwindled. In contrast, northern alfisol soils lost relatively few nutrients to leaching. Supported by crop rotation and nutrient-rich fertilizers, the very nature of northern soils allowed the continuous farming of established agricultural fields.

Soil quality was certainly a major factor in the success of the harvest, especially when cash crops were cultivated for years in the same fields. Traveling in Virginia, Frederick Law Olmsted observed abandoned tobacco fields with "coarse, yellow, sandy soil, bearing scarce anything but pine trees and broom-sedge. In some places, for acres, the pines would not be above five feet high—that was land that had been in cultivation, used up and 'turned out,' not more than six or eight years before."[16] Stunted tree growth revealed low levels of soil nutrients, and the native grass known as broomsedge (*Andropogon virginicus*) frequently colonized these disturbed sites, which eventually became grazing land for cattle. Complex subterranean interactions explain the outcomes. Broomsedge is allelopathic, releasing toxic compounds that inhibited the growth of both competing plants and the nitrogen-fixing bacteria that live in the roots of legumes.[17] Furthermore, Broomsedge provided low-nutrient forage, and thus the manure from cattle that grazed in old fields often offered few minerals when it was used to fertilize crops.

Old fields were also colonized by the native bamboos known as cane (*Arundinaria* spp.). In nature, the largest of these large grasses (*A. gigantea*) formed canebrakes along river banks, clonal vegetation that increased as Native American populations decreased and their fields were abandoned. Its strength and flexibility resulted from elongated cells (sclerenchyma tissue, characterized by cell walls reinforced by additional cellulose and lignin) that accompany the bundles of conducting tissue. This suggested some practical uses; cane strips were woven in chair seats, and the pole-like stems were used in military construction and engineering (see Chapter 8).

Cane, a native bamboo, formed dense riverside canebrakes where travelers and enslaved runaways found shelter; the natural range diminished with overgrazing and land clearing, but populations soon colonized abandoned fields (author's collection).

In song lyrics, Stephen Foster described iconic images of cane in riparian landscapes, including "Down Among the Cane-brakes" in which he described the tall grasses along the Mississippi shore. Travelers camped in canebrakes, and runaways from enslavement hid in the dense cover that could reach twenty feet in height. Cane leaves decomposed and enriched the soil, leading early settlers to use the occurrence canebrakes to identify fertile land for growing corn and cotton. As a result, the natural range of cane diminished during the 1800s as the result of land clearing and overgrazing.[18] As the original riverside canebrakes disappeared, cane dispersed and colonized abandoned fields. Some populations flowered simultaneously (a phenomenon that occurs in other bamboos) and then died. As a result, dead stalks fueled ground fires which cleared competing vegetation and enriched the soil for seed germination and new cane growth; in short, the species tolerated disturbed sites, whether the result of floods, plows, or fires. Cattle grazed on cane, producing manure that was collected and used in cultivated fields.[19]

Overused land was useless to planters and required replacement or rejuvenation. As Olmsted noted, "As their exhausted fields failed to meet the prodigal drafts of their luxury, they only made further clearings in the forest, and 'threw out' ... land as they had ruined. Year after year the process continued ... old-fields, recuperating in the prudent economy of nature, after many years, were again cleared, and, now with some aid of manure, again, for a short time, found capable of producing tobacco."[20] However, plant nutrition was in its early stages as a practical science. Not all micronutrients were entirely understood, and the term *manure* referred to any soil amendment, including animal wastes and mixtures of inorganic minerals. Planters and farmers practiced trial and error soil management, following advice shared by practitioners and theorists. By 1840, Carl Sprengel had articulated his "law of the minimum," his observation that plant growth is limited by the nutrient in the lowest concentration. However, practical experience also contributed to knowledge of soil enrichment for successful crops. Agricultural journals brimmed with advice columns and planters' accounts of success and failure, along with advertisements for proprietary "commercial manures" with unknown chemistry.

An available resource was marlstone, the calcium-rich sedimentary rock formed from clay and fossilized invertebrate shells. As an early source of agricultural lime, pulverized marlstone increased soil pH and provided the calcium needed for the growth of plant cell walls. It was certainly not the only source of soil calcium, thrifty or desperate farmers used pulverized plaster or gypsum with the same results, but marlstone was advertised as promoting remarkable crops in southern clay soils. Rock was mined from deposits in New Jersey and South Carolina and became valued for its agricultural potential. By all accounts, the mineral was highly variable and occurred in various locales; an article in the *Southern Cultivator* noted, "Marl is much used for manure.... There are several kinds of marl—the lime, chalk, loam, clay, and sandy ... from 22 to 32 per cent of calcareous matter...[in] all colors, from black to white."[21] The report of the Centennial Exhibition in 1876 listed 44 different marlstone types by location and chemistry.[22]

There was considerable confusion about marl chemistry, with some dubious claims that the fossil deposits contained high levels of nitrogen and potassium. However, calcium remained the major reason for its application because its ions were needed to raise successful cotton crops on acidic soils. Accounts published in the *Southern Cultivator* and other farm journals claimed the doubling of harvests from crops raised on marled soils, although some growers maintained that the effect of adding agricultural lime was a short-lived, with crops dwindling after a few seasons of spectacular growth.[23] Soil chemistry explains this observation; in areas where calcium was a limiting factor, the addition of limestone to agricultural soils promoted untrammeled growth, which quickly diminished other soil nutrients.

Somewhat confusing were the proprietary marlstone mixtures that often included phosphates and magnesium, a trace element essential for chlorophyll synthesis. Generally speaking, there was poor understanding of the relative importance of nitrogen, potassium, and phosphorus, the three elements in most modern fertilizers. For instance, during the 1860s Carmichael and Bean of Augusta, Georgia, promoted the "National Fertilizer—a preparation of which Marl is the basis … more long lasting that Guano." In fact, guano provided nitrogen, in comparison to the rich calcium content of marlstone. Nitrogen was essential for protein synthesis, while calcium was needed for cell walls, two quite different aspects of plant growth. Marl was also known to New England farmers who dealt with acidic, nutrient-poor granitic soils, and southern growers might have learned from their practices. One practical account described fertilizer production at the state insane asylum in New Hampshire, in which a mixture of marl, clay, muck, feces, and wash water were combined in a massive tank to make a nutrient-laced slurry with "high value as a manure."[24]

Guano

Animal manures included guano, the vast deposits of bird dung that explorer Alexander von Humboldt observed during his travels in South America. It was not really a discovery; guano had long been used by Inca farmers who valued and protected the seabirds that produced it. In 1804 von Humboldt returned to Europe with guano collections for chemical analysis, findings that revealed high percentages of nitrogen, phosphorus, and potassium, the three micronutrients needed most commonly to promote plant growth. Like marlstone, guano was promoted as a quick fix for neglected soils; the nutrients in bird dung promised much improved harvests for European and American cash crops.

The guano industry burgeoned in Peru, with American and European companies mining thousands of tons of accumulated dung and sending shiploads back home for sale. The Peruvian government nationalized supplies and increased prices, and competition for guano supplies became intense. Guano became an expensive soil amendment, but nowhere was it more prized than by southern growers who cultivated cash crops in the same fields with neither rotation nor periods of fallow growth. In his 1850 State of the Union address, President Millard Fillmore observed, "Peruvian guano has become so desirable an article to the agricultural interest of the United States that it is the duty of the Government to employ all the means properly in its power for the purpose of causing that article to be imported into the country at a reasonable price. Nothing will be omitted on my part toward accomplishing this desirable end."[25] By 1850, prices had reached $76 per ton. Some unscrupulous sellers offered guano that was stretched with sawdust, pulverized rocks and minerals, and other adulterants that might resemble the genuine product. As South

Peruvian islands were mined for deep deposits of guano, prized by southern growers who relied on bird dung to restore nutrients to depleted plantation soils. Bird diet, climate, and plowing were important factors in determining the effects of guano on agricultural fields (*Harper's Weekly*, 1865).

American islands were mined bare of their accumulated deposits, new sources were investigated worldwide; these included guano-covered bird islands in the Pacific and coastal islands of England, southern Africa, and California.

Advertisements for products such as Reese's Phospho-Peruvian Guano, Kettlewell's Manipulated Guano, Johnston's Island Guano, and Phoenix Guano were pitched to southern growers and included detailed chemical analysis, endorsements, and testimonials. Products promised lump-free, unadulterated bird dung ready to apply to fields of cotton, corn, wheat, tobacco, and other cash crops. Farm journal editors and correspondents debated the nutrient content and relative merit of various products and the timing and quantity of guano applications; thousands of print pages were given over to guano advertisements, claims, and controversies. Of course, actual guano chemistry depended on the bird species, diet, season, and habitat, and environmental variables that affected guano applications included soil chemistry, climate, plowing, and cultivation.

The chemical detail offered to planters cannot be overstated. *Southern Field and Fireside* (a weekly newspaper published in Georgia) translated the guano analysis provided by Justus von Liebig, whose work on soil chemistry was highly regarded. His volume *Chemistry in Its Application to Agriculture and Physiology* (1847) was one of the first reliable explanations of guano and its use in agriculture, but the editor of *Southern Field and Fireside* debated von Liebig's interpretation of sulfur and phosphate-containing compounds in samples collected from the Baker and Jarvis Islands in the Pacific.[26] Controversies raged about the chemical contents and efficacy of various types of bird dung, and various guano-based fertilizers were discussed and field-tested in the South. The "Prize Essay on Commercial Manures" in the February 1861 issue of the *Southern Cultivator* reported results for 26 fertilizers, ten of them guano-based, but directions for guano use were not generally standardized.[27] The typical application rate was 400–600 pounds per acre, so guano was only affordable for growers of profitable cash crops. On the positive side, guano prevented more deforestation; its use to replace soil nutrients encouraged planters to use land that was already cleared, rather than carving new fields from virgin forest. During his fourteen-month tour of southern states, Frederick Law Olmsted observed the use of guano in "rapidly and permanently restoring the fertility of exhausted soils" when applied by "men of wealth and education" without regard for immediate returns in crop success.[28]

Farmers favored guano for sweet potatoes and other vegetables, but often had to resort to cheaper alternatives for fertilizing their food crops. Some farmers compounded "artificial guano" from mixtures of wood ash (oak and hickory were preferred), rich soil, horse manure, and minerals containing phosphate and nitrogen salts.[29] Another inexpensive option was leaf mold, the composted leaves of shade trees. Directions for its preparation suggested layering maple, elm, or sycamore leaves with horse manure, urine, and waste water; turning the mixture regularly; and waiting about two months for the leaves to be reduced to "a fine mould."[30] Anticipating wartime difficulty with guano imports, Francis Peyre Porcher suggested composting or even burning wild vegetation to use as a guano substitute: "Weeds, leaves of trees, and all the succulent plants which grow so abundantly in ditches and waste lands, under hedges and by the roadside, if cut or pulled when in flower, and slightly fermented, furnish from twenty to twenty-five times more manure than straw does. These plants carefully collected furnish to the agriculturist an immense resource for enriching his lands.... The turf that borders fields and highways may be made to answer the same purpose by cutting it up with all the roots and the earth adhering to them, rotting the whole in a heap and carrying the mass upon the field, or what is still better, by burning it and dressing the land with the product of the combustion."[31]

Cash Crops

Southern planters cultivated gigantic monocultures, the crops grown for cash rather than subsistence. These commodities were the linchpin of Southern agriculture, economy, and even social structure. Sugar cane, tobacco, and cotton so invoked images of the South that Stephen Foster included references to these various crops in the lyrics of songs such as "Old Black Joe" and "Oh Boys Carry Me 'Long." One of the earliest cash crops was indigo, leguminous shrubs that yielded a permanent blue dye which became a valuable export (see chapters 1 and 7). Colonial growers experimented with growing indigo, but the crop proved tricky to raise and the dye they produced proved inferior to West Indian imports. Processing required expertise; the dye does not occur in living plant tissue and required some practical chemistry to produce. Fresh leaves were covered with water, where fermentation converted the colorless glycoside indican to the blue dye indigotin. The dye precipitated as a blue sediment that was collected and dried into blocks for export. Processing depended on careful timing as well as the addition of macerated leaves of elm, mallow, or purslane, which chemically hastened the formation of indigotin.

Two species of *Indigofera* were indigenous to the southern U.S. (*I. caroliniana* and *I. leptosepala*) and were used to make dye for export, but they yielded a substandard product when compared to true indigo (*I. tinctoria*). These Asian shrubs had been cultivated for years in the West Indies, but King George's War interrupted trade during the 1740s and provided an opportunity for American growers to cultivate and ship indigo to England. Beginning in 1739, Eliza Lucas Pinckney experimented with indigo as a possible cash crop for her family's plantation near Charleston, South Carolina. Starting with seeds obtained from the West Indies, she selected and cultivated varieties of indigo suitable for the low country climate. Her dye was ideal for export and in high demand for British-made textiles. Her husband published articles in the *Charleston Gazette* encouraging South Carolina growers to experiment with indigo, and many growers planted indigo as a supplemental cash crop on rice plantations. The naturalist William Bartram (son of the colonial botanist John Bartram) experimented with the crop in Florida. The shrubs grew well on land that could not support rice crops; indigo taproots easily penetrated hardpan clay soil and harbored nitrogen-fixing *Rhizobium* bacteria in their roots. As a result, planters in South Carolina, Alabama, Florida, and Louisiana grew and processed successful crops even in areas with nutrient poor soils.

Indigo trade with England had stopped during the Revolutionary War, and by the 1860s, many growers replaced indigo with more lucrative cash crops. The shrubs were still cultivated in gardens or collected from the wild for household use, but during the Civil War there was renewed interest in native indigo for use as both a dye and medicinal plant. Porcher transcribed a note about *Indigofera caroliniana* from the *Charleston Mercury*, "Our country ladies gather wild indigo, and ferment from it a blue powder equal to the commercial indigo, which dyes a beautiful and everlasting blue. A solution of this powder in water is a speedy and certain relief for cramp and asthma."[32] He included several pages of detailed instructions for cultivating and processing true indigo (*I. tinctoria*) for dye, but also noted that a root decoction might be "useful in nephritic diseases" and a possible poison antidote.[33] Indeed Porcher may have envisioned a return to planting indigo as a southern cash crop after the war; perhaps ironically, there was a Civil War market for indigo from West Indian cultivars to dye the blue cloth used for Union uniforms. However, within a few decades natural indigo became obsolete with the discovery and development of aniline dyes in Germany.

Rice (*Oryza sativa*) arrived in America by the 1650s, probably introduced by ships arriving from Madagascar. Low Country growers converted tidal swamps into flooded rice fields by building levees, canals, and floodgates to control tides and water levels; typical fields were about twenty acres in size and surrounded by soil embankments. Agriculture shifted from cotton and sugar to rice cultivation as soil nutrients disappeared and profits waned. By the 1720s, rice was a major cash crop in South Carolina and other coastal states, and thousands of submerged acres yielded millions of pounds of rice, which were exported to England as an important colonial cash crop. A variety with yellow hulls was introduced in 1784 and became the preferred crop; known as golden seed rice, the cultivar had an aromatic endosperm (nutritive tissue for the embryo) much esteemed by both Creole and English cooks.

Olmsted described the process of clearing forests by girdling and burning trees; charred trunks were often left standing as "rueful corpses of the old forests."[34] Water levels in the fields were managed using a system of gates and ditches to control tidal flow. Rice crops required plowing, planting, hoeing, ditching, reaping, and bundling, and large barns housed rice mills in which the grains were threshed using steam power and prepared for shipment. After harvesting, fields were drained and burned over to reduce the remaining stubble to ash, a practice believed to eliminate pests and weeds and which also returned nutrients to the soil. Rice fields were sometimes later reclaimed for crops of corn, cotton, or sugar cane, relying on enslaved labor to drain and replant the land.

Growers sought enslaved people from the west coast of Africa who were familiar with growing rice in wetlands. Coincidentally many (perhaps nearly half) of these Africans were immune to malaria, a blood cell parasite carried by mosquitos that multiplied in the flooded fields. Protection from malaria resulted from carrying a single allele for sickle cell disease; although planters did not understand the genetic basis for malaria resistance, they certainly realized its economic benefits. Olmsted recorded in his notes that "it is dangerous for any but negroes to remain during the night in the vicinity of the swamps or rice-fields."[35]

Sugar cane originated in Papua New Guinea as a perennial grass (*Saccharum officinarum*) with a sucrose-rich sap, and it travelled across the Pacific as canoe cargo. As the epithet *officinarum* suggests, the plant had early uses in compounding medicines and herbal mixtures. Raw sugar cane was favored for its sweet flavor, even before Indians along the Ganges discovered how to refine pure sugar. Selection began with cane cultivation about 10,000 years ago and resulted in the cultivated varieties that Columbus carried to the Caribbean in 1493. Known for surviving in hot climates, the plants have C4 photosynthesis, an adaptation in which carbon dioxide is absorbed at high levels and stored in four-carbon molecules of oxaloacetic acid, which later releases the carbon dioxide into the usual photosynthetic reactions. C4 metabolism works efficiently at high temperatures when a process known as photorespiration can interfere with efficiency, explaining how sugar cane thrives in extreme heat. In dry conditions, the plants conserve water by closing the stomata on their leaves, and photosynthesis can still occur using stored carbon dioxide.

Sugar cane cultivation and sugar production were exceedingly labor intensive. The plants were propagated asexually by digging rows and planting stem sections (sets) that sprouted shoots and adventitious roots. The mature stems (canes) could grow to twelve feet in height, and they were then cut by hand and pressed to yield a high sucrose sap. Boiling concentrated the liquid into molasses, a process that required collecting and cutting large quantities of fuel. Granulated sugar

Sugar cane cultivation and harvesting were labor intensive. Enslaved workers cut the shoots by hand and pressed them to yield a high sucrose sap, which was boiled to concentrate the sugar. Caribbean sugar plantations provided a model for cane cultivation in southern states, which eventually accounted for a quarter of all sugar worldwide (*Cassell's History of the United States*, 1874–77).

was refined from molasses in a process known as striking. The mechanized process of sugar processing in antebellum mills involved the hazards of grinding rollers and boiling kettles.

Enslaved labor fueled Caribbean sugar plantations and provided a model for cane cultivation in southern states. The crop thrived on fertile coastal land, including acreage in the Louisiana Territory, which was purchased from France in 1803. Hundreds of plantations along the Mississippi cultivated sugar cane as their sole cash crop. American-grown sugar began to replace imports from the West Indies and eventually accounted for a quarter of all sugar worldwide. Of course, there were alternative sucrose sources in the natural world; abolitionists encouraged the use of maple sugar and honey as alternatives to sugar cane (see Chapter 4), and other sugar-rich crops were also investigated. In 1838, the Northampton Beet Sugar Company encouraged cultivation of European sugar beets (*Beta vulgaris*) in the rich alluvial soils of the Connecticut Valley. The sugar produced was of reasonable quality, but the crop did not become popular with farmers although interest in sugar beets revived during the Civil War. Midwestern growers experimented with seed stock from France and Germany, but equipment woes interfered with the refining process.

Like sugar cane, tobacco was another luxury crop with ancient origins, heat tolerance, and early medicinal uses. Long before there were European colonies in North America, Native Americans grew a wild tobacco species (*Nicotiana rustica*) for herbal uses and religious rituals. European herbalists learned about tobacco from the plants carried on slave ships returning from the Americas; they used it to treat gout, fevers, ulcers, syphilis, toothache, lice, parasitic worms, plague, and dropsy (the edema associated with heart failure). Tobacco was often known as Peruvian henbane, suggesting the early herbalists sensed a potent alkaloid chemistry related to true henbane (*Hyoscyamus niger*). Both are members of the alkaloid-rich nightshade family (Solanaceae); the active principle in tobacco is nicotine, an alkaloid that occurs in lower concentrations in other nightshades including potatoes and tomatoes. Jamestown colonists cultivated the South American species (*Nicotiana tabacum*), and tobacco became a desirable cash crop across Virginia and Maryland. By the mid-eighteenth century, almost 70 million pounds of tobacco were exported annually to England, mostly for smoking in pipes.

In the years prior to the American Revolution, tobacco was the leading cash crop grown in both Northern and Southern colonies for export to England. Indentured servants did much of the work of planting, cultivating, and curing tobacco in the colonies. However, by the 1860s, enslaved people produced southern tobacco crops and cleared new land when nutrients became depleted from successive seasons of intensive monoculture. Tobacco plants uptake high levels of nitrogen, phosphorus, potassium, calcium, and several other micronutrients, which diminished soil fertility within a few growing seasons. Soil also affected the final product; the variety known as bright (or brightleaf) tobacco actually thrived on sandy, infertile soil and was traditionally cured by heating the leaves over charcoal fires.

In 1862, the Confederate Congress resolved to encourage planters to replace fields of tobacco with much-needed food crops. Many growers, however, ignored this advice until restrictions were placed on cultivating cash crops rather than grains and vegetables. Tobacco cultivation diminished widely as plantations were damaged by military campaigns and the Union blockade of the Mississippi limited exports. Military rations included tobacco, with demand supplied by any remaining southern crops; Confederate soldiers harvested tobacco from surviving fields and often traded supplies with Union troops in return for coffee. At the same time, Northern farmers planted and harvested tobacco as a cash crop, but many, including editors of *The New England Farmer*, questioned whether "tobacco fever" was a risky fad.[36] By the 1860s, almost all Connecticut farmers cultivated tobacco; some devoted all available manure to fertilizing this crop alone and were forced to purchase hay to feed their manure-producing livestock. The crop was frost-sensitive, prone to attack from tobacco worms, and subject to price extremes, in short, "raising this crop is attended with more care and anxiety than all other crops put together."[37] Seedbed preparation, harvesting, curing, and packing of the leaf crop were all labor intensive processes. Many New Englanders objected to cultivating a crop with no redemptive value "which would do no good to the world … a positive injury to many."[38] There was widespread opinion that the fertile alluvial soils of Connecticut Valley farms were better used for foods such as beans and potatoes rather than vice, and many viewed tobacco as an "exhausting crop" that "stupefied the mind" and left soil "in a very barren condition."[39]

Beginning in 1607, Jamestown colonists cultivated the first southern cotton, and its success over the next 250 years depended on the synergy of enslaved labor, suitable climate, and overseas demand. Seed beds for cotton crops required careful preparation, and the plants needed warm temperatures for fifty to eighty days to mature. Among the various southern cash crops, cotton dominated in all areas of the South where it could be grown successfully. Antebellum planters cultivated upland cotton (*Gossypium hirsutum*) as an annual crop, although the wild populations in Mexico and South American are perennial. Growers on southern coastal islands cultivated Sea Island cotton (*G. barbadense*), a related Andean species with soft seed hairs that produced high quality cloth.

In natural habitats, cotton capsules evolved as an adaptation for seed dispersal. Known as bolls, each hardened capsule splits into four or five valves and exposes the fibers attached to the seed coats. Typical bolls contain about ten seeds, and each is covered by up to 20,000 fibers. These vary in length between the shorter fibers known as linters and longer fibers known as staples or lints. Natural variation in boll size, including varieties introduced from Mexico, provided the genetic raw material for the antebellum selection of improved upland cotton varieties. In addition to improved fiber quality and yield, planters sought varieties with decay-resistant seeds, which at times actually rotted inside the bolls prior to picking. Nineteenth-century planters speculated that insect eggs and larvae might be at fault and used camphor (a terpene derived from the wood of *Cinnamomum camphora*) as an insecticide. Soil quality, high rainfall, and mechanical harm were also blamed for seed rot, described by one planter as "a complete gangrene or mortification of the cotton boles [sic] ... a diseased or morbid condition of the seed."[40] Seed rot is now known to be caused by several species of soil fungi, and some varieties were indeed more resistant to fungal decay than others. One such decay-resistant cultivar was Petit Gulf cotton, a variety named for the bend of the Mississippi River where it was first bred and cultivated.

Cotton roots extended four to six feet deep in the soil, but the plants also developed surface roots that required careful cultivation. Agricultural journals advised planters of "the importance of preserving the roots unbroken ... that it relies [on] for the food it assimilates into cotton and lint."[41] In fact the surface roots were essential for the uptake of dissolved mineral nutrients, which directly affected fiber growth. We now also know that diminished levels of soil potassium affected the quality of cotton fibers, a problem that growers likely encountered if they did not practice crop rotation. Soil fertility was thus an ongoing problem; nutrients were rapidly depleted with successive cotton plantings in the same fields. Some economical planters used surplus cotton seeds to fertilize soil, but fields of decomposing seeds released an "offensive effluvium" that vexed many plantation owners who sought to return nutrients to the soils.[42] Agricultural journals published correspondence debating the relative merits of composting cotton seed before adding it to the soil or by spreading it in fields, but more typical soil enrichment included limestone, potash (potassium-containing salts), ground bones, manure, and guano. Potash was produced commercially from hardwood ashes, but most manufacturers were in New York State, and southern supplies dwindled during the war years.

Confederate agricultural journals printed instructions for extracting potash from botanical waste, including oak leaves and wood, beech and elm leaves, corn stalks, and straw. The typical process involved practical chemistry, in which the ashes from burned plants were mixed with slaked lime (calcium hydroxide) and water, followed by evaporation and heating steps to yield potash ready to pulverize and add to the soil.[43] Even after the Confederate government recommended replacing cotton crops with corn, the *Southern Cultivator* published the chemical analysis of cotton ash, which revealed the high levels of potash (30 percent), limestone (24 percent), and phosphate (35 percent) that cotton crops uptake from soil. Wartime advice suggested that cotton planters practice crop rotation and "a thorough system of fertilization" to avoid abandoning old fields and clearing new land.[44]

The agricultural aim of cotton cultivation was efficiency. The long growing season began with planting seed and extended through three pickings of the bolls as they matured and opened. Wild shrubs can grow over twelve feet in height, but cultivating cotton as an annual crop kept the plants at a uniform height for harvesting. Some desirable varieties such as Petit Gulf cotton had fibers that separated easily from the bolls, which enhanced picking rate, yield, and profit. An early limiting factor was the labor-intensive work of manually separating seeds from fibers; during the 1700s some American cotton was processed using the hand-cranked rollers first devised in India, but the process of feeding fibers into the device was still time-consuming. The invention of fully mechanized cotton gins (shortened from "cotton engines") simplified fiber processing and set the stage for a vast southern cotton industry. Eli Whitney patented the best known of these in 1793; his invention relied on a system of wire teeth and screen-like grids to "gin" the cotton into mounds of seed-free fiber, ready for carding and spinning (see Chapter 7).

Huge quantities of seed waste were a by-product of cotton cultivation. Before binding cotton bales with hemp rope, some unscrupulous planters packed seeds in the middle to add weight. Seeking legitimate uses for excess seed, some planters fattened hogs on cotton seeds, but the prepared mush had to be free of fibers which could prove deadly if eaten.[45] Oil extraction was an option that evolved into an industry; cotton seeds were pressed to yield an oil used to manufacture soap, lubricate machinery, burn for illumination, and dress leather. A report in *Scientific American* described a gasworks in Tennessee known to "generate a magnificent gas from a mixture of rosin, cottonseed, and sawdust" for use in illuminating villages and buildings.[46]

The remaining seed coats and endosperm (the nutritive tissue for the embryo) sold for about a penny a pound as seed cake used for animal feed. Cotton seed cake was superior to flaxseed waste in raising livestock, and farmers sought it as a cheap alternative to grains.[47] According to Confederate reports, by 1858 southern plantations produced 90,000,000 gallons of oil and 800 tons of animal feed from cotton seed waste.[48] Planters eagerly purchased the necessary equipment, and demand fueled additional inventions such as cotton seed hullers, which allowed planters to process seeds after removing the outer seed coats. Some inventions were modifications to pre-existing machinery such as Felton's Patent Grist Mill, which was converted to hull cotton seeds with the addition of a simple attachment. The device was widely advertised to cotton growers in Louisiana and elsewhere.[49]

With ginning and processing mechanized, cotton was planted more widely. Cotton crops did triple duty in producing fibers, oil, and feed, and the harvest increased in southern states from 750,000 bales in 1830 to nearly three million bales by 1850, requiring the scaled expansion of enslaved labor. Because of the number of seasonal tasks required, cotton was known as a crop that kept enslaved workers busy, an aspect of cotton culture that planters favored. By 1860, planters cultivated ten acres of cotton for each enslaved person, who on average picked a hundred pounds a day. Abolitionists opposed the use of enslaved labor to grow and process southern cotton, but the fiber had many essential uses, including the manufacture of cloth, thread, and bandages.

Some Northern farmers attempted to grow cotton to replace crops grown in the South, but successful crops required a minimum of two hundred days without frost, which was unlikely or impossible in all but the most southern of the Union states. In 1862, Congress allocated $3000 to purchase seed stock from Tennessee and North Carolina because their varieties seemed more likely to survive than cotton cultivars from the extreme South. States and counties north of 40° latitude were eliminated as possible cotton regions, but crops were cultivated in some unlikely areas including southern Illinois, where some farm families wove their fibers into homespun cloth. This region became the wartime Union cotton capital, with many small Illinois farmers each planting a few acres of cotton. Labor needs were high for plowing, planting, cultivating, and harvesting; pickers had to be paid, but an experienced field hand harvested as much as two hundred pounds a day. Cotton gins were expensive for farmers to acquire, but shared machines were available to process local crops. Yields reached 250 pounds of ginned fibers per acre, on a par with crops cultivated in the South, and Illinois cotton supplied the high wartime demand of mills in New York and other Union states. Other regions were also planted in experimental cotton crops; in particular, the southwest region of the Utah Territory became known as Little Dixie with cotton yields of 350 pounds per acre.

Self-Sufficiency

Cash crops grown for export lost value in wartime, at the same time that food shortages prevailed and demand for edible crops increased. Trade was disrupted, and so planters often lacked a ready market for much of what they grew. The Confederacy offered bonds that could be purchased with crops rather than money, a scheme that resulted in government-owned stockpiles of cash crops such as cotton and tobacco. However, there was no means to sell these accumulated stores to European countries for the cash needed to wage war.

War dictated a shift in agriculture in which southern growers were encouraged to "plant as little Cotton as possible, for you will have an indifferent market for it … for the sake of your country, that is about to pass through a severe ordeal, make bread and meat!"[50] Headlines such as "Bread for the South!" reminded all that corn and other food crops promised independence from northern control.[51] The Confederate government encouraged replacing cotton, tobacco, and sugar cane crops with corn, wheat, potatoes, sorghum, alfalfa, and other food and forage crops. In 1862, the Confederate Congress passed a resolution suggesting that cotton fields be replaced by fields of corn, and they considered prohibiting cotton cultivation. When growers failed to comply, editors of the *Southern Cultivator* suggested that "the patriotism of the man who contemplates planting cotton next year, in anything like a full quantity, may be questioned, to apply no harsher term."[52] It was generally believed, however, that some wartime cash crops were necessary to provide seed stock for post-war cultivation.

Fields were planted successively with a variety of basic crops. Agricultural journals advised the midsummer planting of cowpeas and pumpkins in corn fields; all three crops provided both food and fodder. In addition to corn ears, the stalks, leaves, and cobs were milled into cheap animal feed, so no part of the corn crop was wasted. Cowpeas (also known as crowder peas, *Vigna unguiculata*) were introduced by enslaved people as an edible legume which became accepted as a table food during the Civil War. The seeds (now known to contain about 26 percent protein) were particularly valued as fodder. Horses fed a diet of half cowpeas by weight were "healthier, sprightlier, able to do more work" that those fed corn alone, which provided a diet of largely carbohydrates and cellulose.[53] In light of the need for military horses, cowpeas became a valued fodder crop.

The *Southern Cultivator* extolled European turnips (*Brassica rapa*) as excellent winter feed for livestock, following English tradition.[54] Varieties of turnips and rutabagas (a turnip-cabbage hybrid) were widely advertised for late summer and fall planting even on newly cleared land. Settlers valued turnips for their versatility. Within a few weeks of sowing the crop, leafy shoots could be harvested as greens for family meals, and the robust mature taproots provided food for both animals and humans. Northern farmers also cultivated rutabagas as animal feed; a Massachusetts farmer advised readers of *The Country Gentleman* to recycle soil fertility by feeding rutabagas to livestock and use their manure to fertilize rutabaga fields, the epitome of New England thrift.[55] By 1863, with food shortages on the Confederate horizon, the *Southern Cultivator* recommended turnips for both human and animal consumption: "Turnips should be sown very largely this year. They are very easily raised, and are good food for man and beast."[56]

Other root crops served as both fodder and food. As a Native American root crop closely related to sunflowers, the tubers of Jerusalem artichokes (*Helianthus tuberosus*) were planted as fresh fodder for livestock. Pigs learned to root for the tubers which thrived and cloned in various types of soil, but some farmers soon realized that the species was potentially invasive.[57] Individual plants can form up to 200 tubers, which efficiently colonized fields if left unchecked. The coarse beets (*Beta vulgaris*) known as mangels, mangolds, or mangelwurzel grew as large as thirty pounds. New England farmers grew mangolds in fields with carrots and turnips, but some noted that the beets often tolerated clay soils better than other root crops.[58] They were grown widely to feed cows and other livestock but were also cooked into soups and stews during times of scarcity.

Sorghum (known as Chinese sugar cane, *Sorghum bicolor*) was particularly desirable because it yielded a caloric sweet syrup that replaced molasses from sugar cane. The crop originated as an African grass introduced to U.S. farmers during the 1850s; William R. Prince of Flushing, New York, recorded his experimental trials with sorghum beginning in 1853, and the crop received attention in agricultural journals. The *American Agriculturalist* mailed eight tons of seed stock to subscribers, reaching 31,000 homesteads; many farmers sowed one or two acres but soon realized that the semi-tropical grass is damaged by early or late frosts. Reports suggested that the quality of sorghum sugar compared favorably to the sugar cane cultivated in Louisiana.[59] However, sorghum yielded better syrup than crystalline sugar, with the potential to replace molasses if grown successfully.

Like sugar cane, sorghum stalks required cutting and crushing to release the liquid that was boiled down into a sugary syrup. There were optimistic claims of producing two or three hundred gallons of syrup per acre, but each vascular bundle in sorghum stems is surrounded by tough sclerenchyma tissue, which made cutting and crushing arduous. Local mills provided the labor and machinery to process sorghum, and these became commonplace as cultivation expanded. Many farmers became disillusioned with fraudulent claims and poor seed stock when some varieties resulted in poor growth, acidic syrups, or other disappointments. Ultimately the Department of Agriculture obtained seed stock from China to replace the inferior cultivars often marketed to American farmers. Some southern farmers repurposed sorghum waste (known as bagasse) as fodder, but there were complaints that the crushed tissue "operates like broken glass, producing violent inflammation" and even death when fed to livestock.[60] Like other grasses, sorghum stems have tough silica-containing cells in the epidermis, and some of the proteins and carbohydrates in mature stems were also difficult for livestock to digest. As a tropical crop, sorghum did not grow aggressively in cool climates, although some farmers in northern states did attempt to make sorghum syrup and the young stems were valued as fodder.[61]

By the 1840s, some Southern planters grew a related species known as Johnson grass (*Sorghum halepense*, native to Asia and northern Africa) as a forage crop. Named for Colonel William Johnson who cultivated the crop on his Alabama plantation, fields sown in Johnson grass yielded large quantities of hay. Seed was accidentally introduced with other crops, and during the early 1800s Johnson grass naturalized in several states. As a close sorghum relative, Johnson grass easily crossed with sorghum cultivated for syrup production, resulting in interspecific hybrids that were useless as a sugar source. In addition to growing from seed, Johnson grass reproduced asexually from rhizome fragments, which sprouted shoots even after a field had been cleared and plowed. Fields planted with Johnson grass were permanently colonized by this aggressive crop, now considered one of the most noxious weeds in the U.S.

Military needs prevailed when the Confederate government imposed in-kind taxes on anything that might be useful to the war effort, particularly the feeding of troops. Ultimately this included cotton, corn and other grains, sugar, potatoes, and various legumes. Subsistence farmers suffered as one-tenth of the food cultivated (and often more) was confiscated for military mess. By the middle of the war years, many farmers began to hoard food supplies in fear of shortages or even famine; many viewed Confederate currency as worthless remuneration for crops and other goods.

Labor shortages ensued as men entered the military. Whether married to Northern farmers or Southern planters, few mid-nineteenth-century wives did heavy labor. Yet work had to be performed, much of it back-breaking when demanded on a large scale. Union spring offensives resulted in the call-up of southern militias during planting season, so agricultural work was left for those at home to perform. Some had anticipated that southern women, many of whom had spent little time outdoors, would become willing wartime farmers; leading up to the war, southern farm journals encouraged women to study crops and cultivation, with appealing references to English noblewomen who were knowledgeable about the land and reminders that George Washington preached a message of agricultural "discovery and improvement" in his last address to the Congress.[62] There were nevertheless concerns that women and girls who worked in the fields would risk injuries, coarsen physically and mentally, and bear social discrimination.

In Chickasaw County, Mississippi, young women drilled as a militia to protect their homes and "to see that the farms are properly cultivated, and full crops raised not only for the support of the county, but the army of Mississippi."[63] However, most southern women learned by trial and error when faced with the responsibilities of planting, cultivating, and harvesting crops. They learned to manage the household labor force of children, youth, and hired men, if they could afford to pay older farmhands. As the war progressed, survival of family and farm pivoted on female labor and ingenuity, despite hardships. Advice arrived in the letters sent by husbands in the army, which could be heeded or ignored, depending on conditions and necessity. Agricultural journals encouraged Confederate farming with glowing descriptions of the inexperienced women who managed to produce crops: "On the small farms throughout this section, all is life, activity, and industry. Many a woman who never before held a plow is now seen in the cornfield.... Many a Ruth, as of old, is seen today binding and gleaning in the wheat-fields."[64]

Fence repair and brush-cutting were ongoing tasks, but other chores reflected the botanical nature of the crops cultivated. Root vegetables such as potatoes, carrots, and rutabagas had to be dug or pulled, which was hard work in dense clay soils. There were heavy field pumpkins to collect in wagons and corn to harvest and husk by hand, an arduous task when done on a large scale. Grain needed scything and arranging in shocks for drying. Hoeing killed weeds that germinated when their seeds were exposed to the light; the more that fields were plowed, the more that they needed hoeing. Yet soil remained the key to success. Agricultural journals reminded wartime farmers of the laborious practices of plowing deeply and overturning clay soils with a spade or mattock.[65]

Fertility of southern soils relied largely on manure provided by livestock, which replaced commercial guano supplies during the war. Some contrarian planters continued to plant cotton on their most fertile land and to dress the crop with manure that was better used on corn fields. Their worst land was often sown in corn, with harvests so minimal that there was little to supply the demands of military taxation. Practical advice for improving soil fertility abounded. Recommendations included collecting swamp vegetation or dead leaves that had been used as animal bedding; this decomposed plant material was then spread on field crops as a source of nitrogen and other soil nutrients.[66] Productivity nevertheless plummeted in the South during the war years, with corn yields falling from fifteen bushels per acre in 1862 to less than half that by 1865. Many fields reverted to tangles of weeds and briars as civilians faced possible famine.

Northern farmers used both manure and night soil, and some developed ingenious methods for processing the human waste collected from chamber pots and privies. One method involved combining night soil with coal ashes and powdered plaster, while another process composted waste with swamp muck and sawdust.[67] A similar commercial product known as poudrette was widely advertised for use in wartime farm fields and vegetable gardens. Lodi Manufacturing and other companies compounded the mixture from night soil (collected from cities), fish bones, and peat; supplies sold readily in New York, Connecticut, and the upper southern states, As early as the 1840s, poudrette was widely advertised in agricultural journals for use on lawns, tobacco, corn, potatoes, and other vegetable crops.

Cash crops were largely retrenched as a source of wartime income, but Southern planters still felt the loss of enslaved labor. Some freed workers became sharecroppers, but the relationship with former owners was frequently contentious. Captured plantations were often subdivided and turned over to tenants who farmed for subsistence crops, although not always successfully. In contrast, Northern farmers were still able to export flour to England (nearly 40 percent of the total crop in 1863), so wartime farming in Union states yielded both food and income. Much of the work of planting and harvesting successful wheat crops was performed by inexperienced New England farm wives who learned practical agriculture firsthand.

Of course, paid labor required remuneration, especially problematic in the South as prices for basic foods inflated far above prewar budgets. Cabbage sold for a dollar a head. Some foraged for the blackberries which grew along fences where birds perched and deposited droppings (see Chapter 4), and even early in the war, these could be sold for 75 cents per quart. Machinery posed another challenge. Women and youth learned to operate various devices used in harrowing, plowing, mowing, harvesting, reaping, threshing, and other tasks, which compensated for some of the labor shortfall. Injuries occurred, but the alternatives were few. Women abandoned corsets and hoop skirts for bonnets, bloomers, and heavy boots suitable for working in fields near dangerous tines and wheels.

The Union military strategy of *chevauchée* raids (see Chapter 2) aimed squarely at Southern agriculture, with the destruction of barns, farm implements, machinery, orchards, warehouses, granaries, hay ricks, and anything else related to crop cultivation and storage. Bivouacked troops allowed their horses to graze in grain fields, which were sometimes intentionally pillaged and trampled. Levees were breached and sluice gates damaged, which flooded and killed crops growing in the fields. As a staple crop, corn became a target; the crop had multiple uses, from cornmeal and fodder to distillation, and became increasingly important in the diets of both civilians and Confederate troops. Stores of the grain were captured, and corn mills were routinely burned. The military goal was to prevent future planting and harvesting, and desperate Southerners eventually traded scrap iron for flour, potatoes, and sorghum syrup. By 1864, farm losses in the South resulted in drastic food shortages, and the Confederate army was in desperate need of vegetables. Townspeople were encouraged to plant gardens on whatever land was available, and small

***Chevauchée* raids by Union troops targeted southern farms and agriculture. Fields were intentionally pillaged and buildings burned, reducing many Confederate families to poverty and subsistence farming (*Harper's Weekly*, 1864).**

crops of root vegetables, squash, cabbages, and potatoes were cultivated wherever they could fit.

The wartime need for hay cannot be overstated. Quartermasters ordered vast quantities to supply military horses, but quality varied. Crops were often still damp when baled, and rocks and logs were included to increase weight. In some areas of New England, 80 percent of the land was planted in hay, and mowing machines replaced the labor-intensive hand-scything of the pre-war years. However, farmers were warned about returning nutrients to the soil, suggesting that "The team that carries hay may bring back a load of manure."[68] Night soil, seaweed, muck, peat, fish waste, limestone, and ashes were also spread on hay fields, depending entirely upon labor and supply.

In the North, there was nothing comparable to the *chevauchée* raids and the deprivation and desperation that they caused. There were regional problems such as insect infections and drought, but the farms in Union states continued intact, with no military campaign to destroy them. On average, prices for many crops rose during the war, often the result of military demand for food, fodder, and fibers. Northern farmers used profits to invest in land, improvements, and machinery. Indeed patent applications for new farm machinery doubled between 1862 and 1865, reflecting the many innovations that were available to farmers with money.

4

Foods and Diet

Crops and Cookery

Antebellum America embraced four major food traditions, each with elements drawn from English ancestry. In the New England states, religious piety was linked with plain boiled dinners and home-baked breads and pies. Local farmers raised the grains, root crops, legumes, and pumpkins that were the botanical backbone of the New England diet. Quakerism influenced foods in the Middle Atlantic region, where dishes were sometimes boiled and bland, or alternately seasoned and southern in flavor. In frontier regions, meals were made from ingredients that could be foraged or stored, and they included dishes prepared from wild greens and cornmeal; indeed English colonists would have regarded much of the American frontier diet as fodder. In contrast, southern food traditions were by far the most diverse, reflecting both African and British colonial influence. As a result, antebellum southern cookery depended on a complex botanical array of vegetables, fruits, herbs, and spices.

Grain is the botanical backbone of all diets, providing both calories and vitamins when the seeds are ground into flour or meal. Botanically speaking, each seed of grain results from two fertilizations, one that produces the embryo and a second that yields the endosperm. Growth of the embryo into a seedling is fueled by carbohydrates and oils stored in the endosperm, which doubles as the nutritive tissue in edible grains. There were no successful North American wheat crops until the late 1700s, so both Northern and Southern colonial cookery depended on Native American corn (*Zea mays*). Known as Indian corn, crops grew reliably in both cool and warm latitudes and had a variety of uses. The kernels were ground into coarse meal or fine flour, dried whole, or converted into hominy by softening the outer layers in hot lye made from wood ashes (see Chapter 8).

As the staple grain, corn was baked into bread, simmered into mush, cooked with beans, and mixed into puddings. Both corn flour and cornmeal were used abundantly, but many preferred wheat flour, which was expensive until the early 1800s. Successful wheat crops coincided with the use of chemical leavening in place of yeast and the invention of wood-fueled stoves with baking ovens. The culinary intersection of wheat flour, baking powder, and reliable ovens resulted in a shift toward baking with refined flour, the cause of many vitamin deficiencies during and following the Civil War.

By the mid-nineteenth century, cabbage and potatoes were the most familiar vegetables on American tables, especially for working class families. Cabbages (*Brassica oleracea*) were bred into early, late, and winter varieties, and they contained sufficient vitamin C to prevent scurvy. Southern farmers overwintered plants to set outside for an early cabbage crop, but some had trouble with having the plants form a firm head. James J.H. Gregory of Marblehead, Massachusetts, advertised the cultivar known as 'Mason,' which he promised was "remarkably reliable in this respect" while also being tender and sweet. It originated as one of the so-called drumhead cabbages grown by Marblehead resident John Mason from a single packet of genetically diverse seeds obtained from a London garden market.[1] In preparation for war, many Southerners grew their own vegetables for the first time, encouraged by Confederate propaganda that encouraged self-sufficiency, and coleworts (cabbages and their relatives) were among the most productive crops. The young shoots known as greens were grown from several species, including colza (rapeseed, *Brassica napus*) which produced a crop in about four weeks. Turnip greens and their robust taproots were also standard fare, usually boiled for long periods in water flavored with pork. Other vegetables were prepared in a similar manner, including cabbage and collard greens (both varieties of *Brassica oleracea*), field peas (*Pisum sativum*), and cowpeas (*Vigna unguiculata*, see Chapter 3).

Irish potatoes (also known as white potatoes, *Solanum tuberosum*) originated in South America and thrived in the warm southern climate. By the 1860s, Confederate agricultural journals included advice on cultivating potatoes, an effort to replace the supplies imported from Nova Scotia for antebellum kitchens. Potatoes became known as a cheap and easy crop that produced tubers even in poor soils, although yield was in direct proportion to the labor invested. In anticipation of wartime needs, a practical article in the *Southern Cultivator* described the process of planting small potatoes in furrows and then mounding soil around the growing plants to encourage tuber formation.[2] Some farmers also experimented with planting the sprawling vines in corn fields, which slowed weed growth and the need for continual soil cultivation. Crops were stored in log buildings with the cracks daubed with clay soil, sufficient to prevent freezing in southern climates. On the lookout for crops with multiple uses, farmers scythed potato vines to use as fodder. There are high levels of the toxic alkaloid solanine in potato leaves, stems, and leaves, but ruminants tolerate levels that cause neurological effects and gastrointestinal upset in humans.

Antebellum agricultural journals reminded farmers to cultivate home gardens of diverse vegetables such as endive, asparagus, celery, and cauliflower; the advice was dubious, however, because all were labor-intensive crops.[3] During the war years, southern families depended increasingly on staple crops such as corn and potatoes as the backbone of their diet. Crops grown for winter use included New World pumpkins (*Cucurbita pepo*), which somewhat resembled the large English squash known as marrows. The fruit type is a pepo, with a tough rind and edible flesh characteristic of the gourd family (Cucurbitaceae). Stored pumpkins kept fresh for months if they were kept cool and dry, or alternately they were sliced and dried like apples. Cooked pumpkin was traditionally used in pies, but the *Confederate Receipt Book* suggested mixing the pulp into bread dough.[4] In *The Virginia*

House-wife (1824), Mary Randolph described an antebellum pumpkin pie made with butter, brandy, and imported spices, but wartime versions were made with whatever milk or eggs were available and flavored with molasses.[5] During desperate times, no parts were wasted. The rind was made into preserves by boiling slices in sugar syrup until transparent, and it was considered a good substitute for citron in cakes. Pumpkin seeds boiled in water yielded "a cooling and nutritive milk."[6] Field pumpkins were long used as fodder, and in times of drastic shortages, kitchen pumpkins were fed to the few surviving pigs and milk cows.

Tomatoes (*Lycopersicon esculentum*) were widely cultivated by the 1860s, and the sprawling vines produced prolific annual crops in southern climates. Native to the Andean region of Peru, tomatoes were introduced to America by explorers, immigrants, and, possibly, enslaved people. Like Irish potatoes, tomatoes are nightshades (family Solanaceae) that were suspected of alkaloid toxicity; the vines do contain solanine and solanidine, but the fruit of cultivated varieties contain only low levels of the toxins. Yet some southern gardeners raised tomatoes as botanical curiosities rather than as an edible crop.

The popularity of tomatoes as a food crop originated with medicinal claims. In 1835, Dr. John Cook Bennett wrote an essay recommending tomatoes as an edible cure for dyspepsia and cholera, and tomato-based liver pills became a faddish alternative to calomel.[7] Bennett's essay was reprinted in several newspapers, including the *Southern Agriculturalist*, and his claims persuaded many Southerners to experiment with cultivating and cooking tomatoes. By the 1860s, there were numerous open-pollinated cultivars that varied in size, shape, and color, ranging from white to yellow, pink, red, and purple. In response to the Confederate need for homegrown food, gardeners started seedlings for early tomato crops in hotbeds; each glass enclosure was filled with several inches of manure that generated heat through bacterial metabolism. Another method promised a "crop that will astonish the natives" by sowing seeds in hollowed-out turnips, transplanting the seedlings into prepared beds, and using wooden frames to keep the developing fruit off the soil.[8]

Fresh tomatoes were preserved by dipping them into a mixture of melted rosin, tallow, and beeswax, which sealed the fruit and discouraged microbial growth, or alternately they were preserved and bottled using molasses syrup.[9] Southern cooks were adventuresome in adopting tomatoes into regional cookery, while New England cookery erred on the side of safety, with long stewing and boiling times for any tomato dishes. In the South, many traditional recipes combined tomatoes and okra (*Abelmoschus esculentus*), a crop with probable origins in Africa or Asia which arrived from the West Indies with the slave trade. Okra capsules were known to release a mucilage that thickened liquids in cookery. Thomas Jefferson grew "tomatas," which his wife and daughters used in preparing gumbos and soups. In *The Virginia House-wife* (1824), Mary Randolph (a distant cousin who was raised by Jefferson's parents) provided recipes for "tomata" catsup, marmalade, and soy, a fermented mixture of tomatoes, cayenne, cloves, and other spices.[10] The *Confederate Receipt Book* described catsup made from vinegar and tomatoes, flavored with mace, allspice, mustard, and black pepper; the mixture was so acidic that it did not require bottling and sealing.[11]

Spicy foods ran counter to New England tradition. Northern cookbook authors such as Catharine Beecher, author of *Miss Beecher's Domestic Receipt-Book* (and sister of Harriet Beecher Stowe), regarded spices as stimulants and observed, "English and American soups are often heavy and hot with spices. There are appreciable tastes in them. They burn your mouth with cayenne, or clove, or allspice. You can tell at once what is in them, oftentimes to your sorrow."[12] In contrast, Beecher advocated a bland, mostly vegetable diet that mirrored the teachings of Sylvester Graham, the nineteenth-century proponent of temperance and food reform who attributed heightened sexuality to spice consumption. New England cooks regarded the heavy-handed use of pepper, cloves, cinnamon, nutmeg, and mace as decadent, unhealthy, and typical of English and southern cookery. According to the tenets of nineteenth-century food reform, spices were medicinal rather than culinary. Indeed spices do have antibiotic properties, a concept that was poorly understood before bacterial disease and food spoilage were understood. In contrast, some southern tastes ran to highly flavored curries and gumbos, food traditions that possibly evolved as an antibiotic defense against food-borne bacteria.

Fruits and Sweets

Agricultural journals encouraged plantation owners and farmers to plant orchards as part of the antebellum aspiration for southern independence. The 1860 volume of *The American Cotton Planter and Soil of the South* included "Essay on the Cultivation of Apples," "Culture of Fruit Trees," and "Benefits of Fruit as a Diet," just a few of the many articles aimed at encouraging fruit cultivation for home use. Some growers many have been tempted to plant fruit trees in depleted fields, but there were pointed warnings about the folly of trying to cultivate orchards on poor soil, such as this advice from *The American Cotton Planter*: "Do not select your poorest, most worn out spot for an Orchard. A person who does not think enough of fruit, to devote a piece of *good* land to his fruit trees, to plant them properly, and to cultivate them well, had better not plant any at all."[13] However, growers soon learned that peach trees in fact tolerated poor soil, while apple and pear trees grew best on soil not depleted by the cultivation of cash crops.

Peaches, apples, and grapes were among the most popular fruit, consumed fresh in season or dried for winter use, but quince, figs, pomegranate, and pears were also widely cultivated in home gardens and orchards. Articles and advertisements in agricultural journals reflected the timely interest in fruit cultivation; the trees were described as growing to their maximum size and productivity in southern climates, but orchards took years to establish. As noted in the *Southern Cultivator*, "As to Fruit, no country on the globe is so well adapted to its production ... a climate suited, in its broad range, alike to the Orange and the Apple, the Olive and the Pear, the Date and the Grape."[14] Some orchardists even argued that fruit was more profitable than typical field crops and that surplus could be sold to northern states.

Cultivated apples (*Malus pumila*) originated in Europe but grew more abundantly and self-seeded readily in New World habitats. They were the most common cultivated fruit in both the North and South, and they assumed a quasi-medicinal role in diet as a natural alternative to cakes and other sweets. Across nineteenth-century America, constipation was a national scourge caused by the dietary preference for meat and starch. An article in *The Ladies' Home Magazine* advised readers to keep several barrels of apples on hand as a food useful in "removing constipation, correcting acidities, and cooling off febrile conditions, more effectively than the most approved medicines."[15] Apples were also needed for making cider vinegar which was used as preservative in pickling and as a household remedy (see Chapter 5).

The southern climate encouraged the cultivation of numerous early, summer, and fall apple varieties with cultivar names such as 'Disharoon,' 'Oconee Greening,' 'Green Crank,' and 'Stevenson's Winter.' Vast numbers of southern varieties were selected or developed for other fruit crops; nurseries such as Fruitland in Augusta, Georgia, advertised dwarf and standard trees, including over 200 peach (*Prunus persica*) cultivars such as 'Amelia,' 'Nutmeg,' 'Druid Hill,' and 'Stump the World.' 'A Pear List for the South' in *The American Cotton Planter and Soil of the South* recommended 'Madeline,' 'Beurre Giffart,' and 'Jargonelle' as fine pears for southern climates.[16]

Orchardists selected desirable fruit varieties with unique flavors, pigments, and seasonal growth. Hybridization was poorly understood, but growers realized that seedlings often did not produce fruit similar to the parental stock and indeed that if "artificial or accidental fecundation takes place ... by insects, by wind or other causes, then it is almost certain that individuals dissimilar to the parents will be obtained."[17] The principles of heredity were not yet known to American horticulturalists; Gregor Mendel published his work on plant hybridization and heredity in 1866, but his paper was not widely read until 1900. Thus observation was key to identifying individual trees that produced well in southern climates, and these were propagated by cuttings.

Grafting was another option, in which "apple and pear trees that produce inferior fruit, may now be grafted in their limbs with choicer varieties."[18] This technique allowed farmers to plant disease-resistant trees and then graft desirable cultivars as branches on the sturdy rootstock. Debate continued in wartime agricultural journals about the likelihood of seedlings resembling the parent, which was unlikely if cross-pollination occurred with other varieties of the same species. In contrast, asexual propagation by cuttings and grafting produced reliable results. However, inconsistent propagation meant that many southern fruit cultivars vanished after the war, although some may have been similar or even identical to nursery stock sold in the North.

Fruit may have become scarce during the war years, even in northern kitchens, but wartime volumes of *Godey's Lady's Book and Magazine* recommended many apple dishes in their seasonal menus, including apple soufflé pudding, apple sauce, pork pie with apples, hot apple tart, and apple tartlets made with green apples. Dried apples were used in making pies and dumplings. If supplies ran low, the *Confederate Receipt Book* included a recipe for "Apple Pie Without Apples" prepared from crackers flavored with sugar, nutmeg, and tartaric acid, a crystalline acid which occurs naturally in grapes and provided the slightly tart taste typical of fruit pies.[19]

Labor shortages resulted in neglected crops, and the military strategy of *chevauchée* raids (see Chapter 2) took aim at Southern agriculture, including the orchards that supplied fruit for homestead meals. Union soldiers on the march stripped the remaining fruit from branches, and so for many Southerners foraging became the alternative to protecting and maintaining orchards. Foraged wild fruits included barberries (*Berberis vulgaris*), which were introduced by early colonists and widely naturalized; wild cherries (*Prunus virginiana*); and blackberries (the native wild blackberry, *Rubus occidentalis*; highbush blackberry, *R. villosus*; and the southern dewberry, *R. trivialis*). Blackberries were particularly widespread during the war years because they readily colonized abandoned fields; seeds of the thorny shrubs are bird-dispersed and germinated near the fences where fruit-eating birds perched. Blackberries cloned by sprouting new canes from the horizontal root system, and these survived in agricultural fields because nineteenth-century plows did not reach the twelve-inch depth of the roots. Various species of *Rubus* hybridized freely and readily acclimated to depleted agricultural soils.

Blackberries are an aggregate of small stone fruits (known as drupelets, each resembling a small plum or cherry). With high sugar content, they were useful in pie-baking and particularly in making household and medicinal wines, which substituted for wine made from grapes. Wild yeast occur naturally on the fruit epidermis and ferment sugars into alcohol quickly, the basis for the blackberry wine described in the *Confederate Receipt Book* as being so flavorful as to "make lips smack as they never smacked under similar influence before."[20] *Resources of the Southern Fields and Forests* described wine preparation from the crushed berries, brown sugar, nutmeg, and cloves, a recipe identical to one that appeared in issues of *Godey's Lady's Book and Magazine* from the early 1860s.[21] The mixture required a few weeks to ferment into a weakly acidic wine, which was bottled and left to age. More importantly, the berries were a valuable scurvy preventative with almost twice the vitamin C content of tomatoes, particularly useful at a time when meals may have lacked essential nutrients. Practitioners treated dysentery with blackberry cordials and root decoctions, a cure based on the astringent properties associated with high tannin content. Many Southerners picked the berries in abundance during the war years and Reconstruction, a means by which women and children augmented both their farm income and household diet.

Sugar consumption increased during the nineteenth century and included primarily cane sugar and sorghum (see Chapter 3). Molasses, the waste liquid from sugar refining, was used widely as a table syrup and in cookery, especially in the South. It was often purchased by the barrel, as was sorghum syrup when cultivation of sorghum crops expanded. As might be expected, New England cookbook authors and food reformers condemned the over-consumption of sugar. In *The American Woman's Home*, Catharine Beecher and Harriet Beecher Stowe blamed the unhealthy preference for sweets on the ancestral British taste for heavy plum puddings, described as "every indigestible substance you can think of" boiled and served in flaming brandy.[22] The dish contradicted every dietary principle that the Beechers and other reformers advocated.

Many New England social reformers were Abolitionists who avoided any foods or other crops grown using enslaved labor, including sugar cane, rice, and cotton. In Union states, syrup and sugar made from the sap of sugar maple trees (*Acer saccharum*) became a common alternative to plantation-grown cane sugar. Maple sap is a nutrient solution of sugars (mostly disaccharides), proteins, amino acids, vitamins, organic acids, and minerals, transported during the late winter and early spring in phloem, the food conducting tissue beneath the bark. Colonial settlers learned from Native Americans how to collect the sap and concentrate the sugars by boiling, and the process became a tradition in northern states. By 1860, the Currier and Ives catalogue included "American Forest Scene," an engraving based on a painting by A.F. Tait of maple sugaring operations in a New England forest; local people carry sap buckets, tend the fire, and boil sap down to a fraction of its original volume. The engraving illustrates the Abolitionist message: Maple sugar can replace slavery-dependent cane sugar. Nor was this the first time that New Englanders produced maple sugar as an act of defiance. A century earlier, patriots tapped maples in response to the Sugar Act of 1764, in which the British Parliament taxed cane sugar imported to the American colonies. During the Civil War years, a series of paintings by Eastman Johnson further reinforced the Abolitionist message by illustrating the wholesome social and technical details of maple sugar production. As a Union supporter, Johnson also followed troops on three campaigns and documented aspects of military and civilian life in oil paintings.[23]

Articles in the *Report of the Commissioner of Agriculture for the Year 1862* described sugaring methods clearly and encouraged experimentation. Maple sap flowed in the late winter when there was little else to do on a family farm, so sugaring became a social event across generations, as noted by the *Report*, "The work of manufacturing sugar and sirup takes place at a season of the year when other active farm labor has not been resumed, and thus affords a good opportunity for performing one of the pleasantest and most social parts of farm work."[24] Yet despite the demand for maple sugar, the mid-nineteenth century was a time of large scale logging in Union states, which destroyed much of the maple-dominated woods known as sugar bush. Maple timber was used for lumber or making potash, the potassium compounds used in making fertilizer, soap, gun powder, and glass. Incidentally, some in the South knew of maple sugaring, which Frances Peyre Porcher described in *Resources of the Southern Fields and Forests*, citing information gleaned from the *Farmer's Encyclopaedia*.[25] He suggested sugaring

Abolitionists promoted the use of sugar made from the sap of sugar maple trees as an alternative to plantation crops cultivated by enslaved labor. Colonial settlers learned from Native Americans how to collect the sap and concentrate the sugars by boiling, and the process became a tradition in northern states (morphart © 123rf.com).

as an option for civilians strapped for food and cash, but apparently did not realize that the natural range of sugar maples extends only as far south as Kentucky and central Tennessee.

Grains and Bread

Historians have long described corn and pork as the backbone of diet in the antebellum South, but corn alone drove the edible economy of the Confederate states. Crops provided flour, meal, fodder, and mash, the crushed grain that was fermented and distilled into whiskey. Native American corn became the staple food in colonial Virginia, where traditional baking during the early nineteenth century relied on recipes such as the cornmeal bread described by Mary Randolph in *The Virginia House-wife* (1824).[26] This was a rich mixture of cornmeal, milk, eggs, butter, and yeast, far from subsistence cookery. Her biscuit recipes also required large quantities of butter and eggs; drop biscuits were leavened by several beaten eggs, and tavern biscuits resembled shortbread. *The Virginia House-wife* included recipes for cornmeal mush and polenta (mush cooked, cooled, sliced, dressed with butter and cheese, and baked) and puddings of "Indian meal" and hominy prepared from corn kernels (see Chapter 8).[27]

A few decades later the recipes for comparable foods were drastically simplified. Cornbread (known as Indian bread) was the daily bread of the antebellum South, even among those who could afford to bake with wheat flour, at least for Sunday meals. Traveling in Virginia, Frederick Law Olmsted described a meal accompanied by corn-cakes and preserved fruits; he recorded various recipes for cornbread, some as simple as baking cornmeal moistened with water alone.[28] In all cases, these were served hot, in contrast to the New England practice of eating day-old bread at room temperature, cut easily into thin slices and considered more digestible. Typical cornbread was a mixture of cornmeal and some sort of flour, sweetened with molasses and leavened with baking soda.[29] Versions of cornbread varied with the addition of whatever milk, eggs, or fat was on hand. *The Confederate Receipt Book* (1863) provided wartime versions of breads, including biscuits made of any flour available and a small amount of fat. Cornmeal and water alone were baked into the unleavened bread known as pone; cornmeal cooked in water or milk was fried into hoecakes.

Various baking powders and baking soda (sodium bicarbonate) replaced yeast in many wartime bread recipes and could be kept on the shelf without the care required by live yeast cultures. Their use began in the late 1700s with the potash or pearl ash that was leached, boiled, and crystallized from wood ashes. Chemically speaking, baking powders and baking soda are alkaline mineral salts that react with acids in the batter to produce carbon dioxide as leavening. The reaction occurred quickly in the intense heat of the iron cookstoves that became popular in the 1860s, and their use spread as wartime stringencies increased. Some recipes specified saleratus (derived from *sal aeratus*, Latin for "aerated salt"), a commercial product that contained potassium carbonate and various additives. Sold in small envelopes printed with images of fancy cakes, saleratus allowed cooks to bake frugally with fewer eggs needed for leavening. The traditional recipes of Mary Randolph and others had required eggs by the dozen, which during the war years may have been in short supply.

Baking with chemical leavening often did not compare well with traditional yeast methods. In *The American Woman's Home*, Catharine Beecher and Harriet Beecher Stowe complained about "the green, clammy, acrid substance called biscuit" with regret that "the daughters of our land have abandoned the old respectable mode of yeast-brewing and bread-raising for this specious substitute."[30] In southern climates, stored flour and cornmeal became rancid or insect-infested, and cooks often had to use whatever was edible and available. Many Southerners preferred the light color of wheat bread and the paler varieties of cornmeal, to the extent of cultivating white rather than yellow corn crops. Faced with shortages of wheat flour, some resorted to dark bread baked from various proportions of cornmeal and rye flour, wartime loaves that resembled the traditional brown bread baked by New England settlers.[31]

Corn was the staple grain in southern diets, used for making bread, hominy, mush, whiskey, and fodder; wartime cornbread recipes called for little more than cornmeal, water, and a bit of fat. Corn prices soared when farmers realized that crops were more valuable for distilling whiskey than baking bread (author's collection).

Rice was the only other grain commonly consumed in the wartime South, primarily in the coastal areas of the Carolinas and Georgia where it was cultivated (see Chapter 3). Bowls of rice replaced cornmeal mush as a breakfast dish, and it was used in baking, such as the method described by Olmsted, "A very delicate breakfast-roll is made in Georgia, by mixing hominy or rice, boiled soft, with rice-flour, and milk, in a stiff batter, to which an egg and salt may be added. It is kept over night in a cool place, and baked … brought hot to the breakfast-table."[32] *The American Cotton Planter and Soil of the South* reminded readers that "everyone should keep rice on hand all the time" and suggested using it to increase the volume of a custard pie made from milk and eggs.[33]

Grain shortages resulted from *chevauchée* raids that destroyed granaries and warehouses (see Chapter 2). Union troops captured stores of grain and burned corn mills, and bivouacked troops allowed their horses to graze in grain fields, which were sometimes intentionally pillaged and trampled. Drought conditions during the summer of 1862 compounded grain shortages in Confederate states, at a time when other foods were already in short supply. Bread prices in southern cities increased as flour supplies dwindled, leading to protests about exorbitant prices during a time of need. Mobs in Virginia, Georgia, and the Carolinas attacked warehouses and markets; in Richmond, the Confederate capital, furious women armed with makeshift weapons looted food and other goods. Purportedly Northerners had a hand in instigating the unrest. Jefferson Davis, the Confederate president, threw money at the crowd of civilians, who refused to disperse until threatened with the local militia.

Violent bread riots erupted in several Confederate states in response to Civil War food shortages, the result of military demand, labor shortages, poor transportation, farm raids, inflation, and myriad other factors. Women struggled to feed their children, and soldiers found it difficult to fight while worrying about food-strapped families at home (*Frank Leslie's Illustrated Newspaper*, 1863).

In response to severe shortages, Civil War food riots (often known as bread riots) resulted from several factors, including military demand, farm labor shortages, poor transportation of goods, destruction of crops and farm machinery, war refugees crowding into cities, and planters' reluctance to replace cash crops with food crops. Inflation further contributed to wartime food insecurity in southern cities, as the cost of war exceeded tax revenues and Confederate currency diminished in value. In short, many women found it impossible to purchase bread and flour needed to feed their children, and soldiers found it difficult to fight while concerned about family survival.

Food Preservation

At a time when cookery depended on dried, pickled, or stored foods, survival required successful techniques for food preservation. This was a new experience for many southern women, who once had depended on enslaved people to prepare family meals and who now found themselves responsible for feeding children, elders, and hired farm hands. Southern agricultural journals certainly promoted the idea of food preservation as a means of independence; with war just two months away, an article on Queen Victoria's family while in residence at Osborne on the Isle of Wight detailed the practical education of her daughters "cooking the vegetables from their own gardens, preserving, pickling ... to distribute to the poor of the neighborhood, the results of their handiwork."[34] If the royal princesses could learn to preserve crops, so could women of the South.

Several techniques were known and used. Meats were preserved by smoking with corn cobs or woods such as apple and hickory, and ashes from green hickory combined with salt were used as a preservative for bacon.[35] Summer and fall harvests required preserving fruits and vegetables for use in the winter and early spring. Dehydration was one simple option; sliced apples or pumpkins and okra capsules were often strung and air-dried. If stored away from moisture and vermin, these dry and inert botanical foods kept indefinitely and were easily rehydrated for pies, soups, and stews. Some fruits and vegetables were stored in deep cellars that kept foods consistently cool, which was particularly important in warm southern climates. Whole fruits and vegetables contain living cells that respire at low rates, and cool storage kept cells dormant. Good air circulation was essential because living tissue needs oxygen and releases carbon dioxide, so fruit were often stored in baskets between layers of paper and straw. Avoiding frost was important, as was air circulation to avoid the accumulation of ethylene, the hydrocarbon gas (C_2H_4) that as a natural plant hormone stimulates over-ripening of stored fruit. The effects were first discovered in city trees; gas streetlights released ethylene, which the trees recognized as a hormonal signal to defoliate and enter dormancy. Ripening fruits release ethylene, which in turn causes changes in color, texture, sugar concentration, and acidity. Bruised plant tissues also release high levels of the hormone, signaling nearby fruit to over-ripen and leading to microbial activity and decay. Ethylene from a single rotting apple spoiled an entire barrel (or basket) of fruit, so care had to be taken to store only unblemished fruit and to inspect them often.

Another familiar method involved packing fruit such as pears between layers of dry sphagnum moss (*Sphagnum* spp.) in earthenware jars, sealing the lids with rosin (solidified conifer resin), and burying the jars in a foot of sand. The moss absorbed moisture, the rosin protected against pests, and the sand kept the jars cool. Alternately fruits and vegetables were buried directly in sand or sawdust; vegetables were hilled up in corners of cellars, and cabbages were stored in pits where they often stayed fresh until spring. Southern farmers believed that cotton delayed ripening and prevented decay, and they used it to pack soft-skinned fruit, grapes, apples, and pears in sealed boxes for cool storage into the winter months.[36] In fact, cotton does not have antiseptic properties, but the fiber-lined boxes prevented damage and excluded insects and air-borne microbes associated with decay.

Fruit preserves, the bottling of cooked fruits in sugar syrup, were part of the British colonial culinary tradition. High sugar concentrations dehydrate and inhibit microbes, and so cooks had to plan on approximately the same weight of fruit and sugar for long term preservation. Some bottling practices during the 1860s reflected thrift, including the recommendation in *Resources of the Southern Fields and Forests* (1863) that preserves be prepared from rinds that might otherwise have been discarded: "The harder portions of both watermelon and pumpkins are used in making preserves by our Southern matrons."[37] Wild fruits such as crabapples (*Malus angustifolia* and related species) and buffaloberry (*Shepherdia argentea*) were tart but suitable for preserving with sugar. Buffaloberries are native to central and western North America, but by the mid-nineteenth century they were naturalized outside of their range; the bright fruit resembled red currants and were adopted for tarts and preserves. Cranberries (*Vaccinium macrocarpon*) were known from some areas of North Carolina, where they were gathered and preserved by drying or by bottling the berries in fresh water.[38]

Vinegar and salt were essential ingredients in food preservation. The South depended on northern states for basic commodities, including the salt used in pickling and meat preservation. Union troops destroyed new coastal saltworks as quickly as they were established, and so southern salt shortages persisted for the duration of the war. Although grains were also in desperate demand, local firms offered to trade a bushel of salt in return for four bushels of corn or two bushels of wheat.[39] In other words, strapped families had to choose between baking bread or preserving meat and other foods by salting or pickling.

Vinegar made from apple cider was widely regarded as the best preservative for pickles. The high acidity (low pH) of vinegar prevented microbial activity, and the flavor of cider vinegar contributed to its value for pickling fruits and vegetables. The process began with wild yeast that fermented sugars into a slightly alcoholic (hard) cider, which was then intentionally inoculated with acetic acid bacteria from vinegar already on hand. Known as the "mother," the cloudy mass of bacteria converts alcohol (ethanol) into acetic acid. A barrel of cider vinegar was often kept for years by adding leftover cider and wine to replenish and feed the culture of acetic acid bacteria. In addition to its use in food preservation, vinegar provided the tart flavoring of non-alcoholic drinks such as shrubs and switchels, often served in hot weather. There were also medicinal uses, and vinegar became a wartime necessity for its antiseptic properties, treating fevers and dysentery, and preparing certain drug mixtures (see Chapter 5).

In addition to salt and vinegar, standard pickling recipes also often called for ginger, turmeric, and cinnamon, which may have been in short supply during the war years. Some cooks may have substituted the powdered bark or leaves of sassafras for cinnamon (both are in the laurel family, Lauraceae). The roots of wild ginger (*Asarum canadense*) resembled the flavor of Asian ginger although the species are unrelated. Beets and cabbage were easily preserved in vinegar, perhaps with the addition of cayenne pepper, and other ordinary pickles were made from olives, mushrooms, cauliflower, and cucumbers. Cookbooks and magazines such as *Godey's Lady's Book and Magazine* included more elaborate methods for pickling fruits, ranging from small melons to the fruit of Brazilian devil's claw plants (*Ibicella lutea* and related species in the martynia family, Martyniaceae). These tropical plants thrived in conditions that favored okra and readily self-seeded, and its yellow catalpa-like flowers were a garden curiosity. The immature capsules (known commonly as martynias or "martinoes") were pickled in spiced vinegar; in nature, the woody capsules catch on animals' feet and are a unique adaptation for seed dispersal.

Recipes for Indian pickles included a variety of fruits and vegetables for preservation in brine, including gherkins, onions, cauliflower, celery, broccoli, and peppers. Peaches and melons were filled with mustard seeds and garlic, reminiscent of the practice of pickling mangoes stuffed with spices.[40] Even New England cookbooks such as *Miss Beecher's Domestic Receipt-Book* recommended recipes for preserving various fruits and vegetables in vinegar, salt, and spices, perhaps because pickles were consumed in small quantities and brightened otherwise dreary meals.[41]

Canning was invented during the early 1800s by Nicholas Appert, a French chef and confectioner who perfected the hermetic seal as part of a competition to provide food supplies for Napoleon's troops. Mason jars for storing preserved fruits and vegetables were patented in 1858 by John Mason, a New Jersey tinsmith. Some nineteenth-century home cooks used Mason jars for preserving fruits and vegetables, but the process was risky for low-acid foods such as pumpkin, corn, and beans. Jars of such foods provide an oxygen-free environment for soil bacteria (*Clostridium botulinum*) that cause botulism, the deadly poisoning from nerve toxins that cause paralysis and halt respiration. A combination of heat and pressure are needed to kill the spores of *C. botulinum*, but pressure cookers for processing canning jars were not available for home use until after World War I. Civil War–era cooks who canned low acid foods had to practice extreme cleanliness to avoid any soil contamination and the risk of deadly botulism toxins.

Military Rations

Diet was part of the ordeal of warfare. For both Union and Confederate troops, meals were often feast or famine, often depending on raided supplies, preserved foods, and packages received from home. Nutrition was poorly understood, so quantity rather than quality defined most Civil War rations; lacking refrigeration and consistent access to botanical foods, troops experienced vitamin deficiencies that affected their health and productivity. Vitamins and their metabolic functions were not discovered until the early 1900s, so conditions such as scurvy and night blindness were not always attributed to deficient diets. The daily ration for Union soldiers included vegetables and flour, but shortages resulted in competition for food and fodder, with corn needed to feed both men and horses. Perhaps the most welcome but modest military meals were provided by battlefield nurses who served wine with pulverized crackers or bread. Nothing was wasted in battlefield settings; the Civil War nurse Clara Barton was remembered for the gruel that she prepared from the cornmeal used to pack crates of medicinal wine.

Rations relied on botanical foods that included wheat and corn flour, cornmeal, potatoes, legumes (dried peas and beans), canned and dehydrated vegetables, rice, and hominy (corn grains with the outer layers softened in lye, see Chapter 8). Onions and dried apples and peaches were infrequent but welcome. Vinegar rations were used to preserve and pickle cabbage and other vegetables, which added a few vitamins to deficient diets. Sugar and molasses added flavor and calories, and coffee and tea (or substitutes) provided stimulation and comfort, but foods exceeding the basic rations required the means for self-sufficiency. Using large-scale methods developed by Appert, Civil War canneries provided food that could be shipped to military camps and taken with troops on the march. Canning methods improved as a result of wartime need, and some foods were preserved in cans for the first time. Union troops consumed canned wild blueberries (*Vaccinium angustifolium*) picked by civilians in Maine.

Some soldiers purchased their own supplies from sutlers, merchants who sold fresh onions, fruit, and other desirable foods directly to men who could pay in cash. However, most troops obtained fresh fruit and vegetables through raids and theft (known ironically as "foraging") at the expense of civilian farmers and plantation owners. Actually foraging for edible foods was another option. Confederate troops searched for wild onions (*Allium canadense*), which some thought had a flavor superior to cultivated onions, although it is now known that ingestion interferes with normal thyroid iodine uptake.[42] Soldiers also foraged onions for their medicinal properties; cultivated onions (*A. cepa*) and related wild species had folk uses as vermifuges and expectorants, and *The Confederate Receipt Book* mentioned using onion poultices to cure infections.[43] In fact, when onion tissue is damaged, a series of chemical reactions converts the sulfur-containing compound alliin into allicin, which is antibiotic to both *Salmonella* and *Staphylococcus*.

Wild blueberries and huckleberries (*Vaccinium* spp. and *Gaylussacia* spp.) were widely collected, and tea was brewed from the leaves and bark. In comparison, the dark fruits of the related species known as farkleberry (*V. arboreum*, also known as sparkleberry) were bitter and tough, but they could be foraged in woodlands, meadows, and creek banks across its wide southeastern range. Unripe persimmons (*Diospyros virginiana*) were notoriously astringent, so soldiers often waited until after the first frost to collect the large berries for consumption; these wild fruit were abundant in many areas of the South and provided a seasonal addition to meager rations.

At the time of the Civil War, American chestnuts were common trees in the eastern U.S., comprising up to half of some hardwood forests. Along with acorns and beechnuts, chestnuts were a significant part of the mast that accumulated on forest floors. They provided an important animal and human food until forest populations of wild trees succumbed to the fungal blight (*Cryphonectria parasitica*) introduced in the early 1900s with imported Chinese chestnut trees. For troops encamped near hardwood forests, chestnuts (*Castanea dentata*) were a particularly useful food to forage; cooked or raw, they were a high-calorie source of protein and fats. Chestnuts, walnuts (*Juglans nigra*), and pecans (*Carya illinoinensis*) were considered preferable to other wild nuts, but in times of need acorns (*Quercus* spp.) and beechnuts (*Fagus grandifolia*) were also collected and eaten raw, roasted, or boiled.

The most iconic Civil War food was hardtack, square crackers invented with longevity rather than palatability in mind. The only ingredients were flour, salt, and water because oil or fat would have caused stored supplies to become rancid over time. Soldiers on both sides depended on hardtack while on the march or encamped in temporary quarters; supplies doubled as a food and also as an ingredient in military cookery. Hardtack provided calories for marching and fighting, but its vitamin content depended on the percentage of the embryo (germ) and bran (outer layers of the grain) used in milling the flour; baked with white rather than whole wheat flour, hardtack contained few vitamins and no protein. Other botanical foods such as legumes and vegetables provided protein and vitamins, but balanced military meals, in a nutritional sense, were infrequent at best. Corn was the backbone of the Confederate military diet, primarily as the source of cornmeal, and there were times when troops had only cornmeal to eat. In addition to cornbread and other traditional ways of preparing cornmeal, *The Confederate Receipt Book* recommended the mixture known as sagamite, cornmeal heated with brown sugar, which supplied quick energy for soldiers on the march.[44]

A corn-based diet resulted in conditions caused by vitamin deficiencies, and so civilians were asked to contribute curative foods to Confederate rations. Civilians planted extra tomatoes, carrots, squashes, cabbage, and other vegetables with known antiscorbutic properties (scurvy was known as scorbutic fever), and these were shipped to local encampments. Francis Peyre Porcher encouraged home cooks to dehydrate tomato pulp to give to the military.[45] As farms were destroyed and shipments blocked, troops experienced definite malnutrition; simultaneously food prices inflated, with potatoes often costing ten times more in Confederate cities that in the North. Confederate soldiers scavenged for wild botanical foods known to combat the symptoms of scurvy, including blackberries, sassafras buds, wild garlic and onions, clover, and the shoots of pokeweed (*Phytolacca americana*, now known to be potentially mutagenic, although folk wisdom has long recommended the young shoots as potherbs). Blackberries in particular augmented their meager rations, and men made cordials and canned preserves from the surfeit.[46]

Civilians shared meager supplies with army hospitals, where sick and wounded Confederate soldiers received gifts of blackberries, peaches, onions, and pickles. Similar shortages occurred in Union hospitals; based on her time as a volunteer nurse with Union troops in Washington, Louisa May Alcott described bread containing sawdust and leavened with saleratus (an early type of baking powder) rather than yeast, weak tea steeped from huckleberry leaves, and plenty of stewed blackberries.[47] Prisoners of war in particular suffered the ill effects of poor diet. Relying on prison rations alone, captured soldiers were the most likely to suffer from the effects of scurvy, which included the loss of hair and teeth, the inability to walk, and eventual death. Diagnosis involved a simple thumb press that revealed discoloration of flesh; botanical foods were the cure, but fresh fruits and vegetables were excluded from prison rations.

Depending on circumstances, military meals were prepared by poorly trained cooks who presided over field kitchens or by individual soldiers who lacked proper equipment and often knew little about cookery. Food was often spoiled or of dubious quality, and army cooking during the Civil War was notoriously poor, even when decent food was available. Packages from home likely contained cookbooks such as *The Cook's Own Book* (1832 and later editions, Monroe and Francis, Boston) by

Mrs. N.K.M. Lee, which many southern cooks owned by the 1860s. Although slanted toward New England food preferences, it was as much a reference text as an alphabetical compendium of recipes, which may explain its popularity. Most importantly, *The Cook's Own Book* described various methods of cookery, including a lengthy section on soup preparation that demystified simmering and skimming. Cookery that stretched meat rations with copious vegetables was key to the preparation of military meals, but it required elements of planning (such as cooking a day ahead) that were revelations to most men. Soldiers complained about the unexpected tartness of fruit pies made without sugar, which was often in short supply and saved for nourishing sick or wounded soldiers.

Union troops used the War Department publication *Camp Fires and Camp Cooking: Culinary Hints for the Soldier*, authored by Captain James M. Sanderson, a former hotel operator. The manual provided plausible instructions for building fires in trenches and suspending iron kettles above the flames, but Sanderson's credibility was diminished by his claim that "No army in the world is so well provided for, in the shape of food, either as to quantity or quality, as the army of the United States."[48] Perhaps the most welcome advice in *Camp Fires and Camp Cooking* centered on how to use wood-burning portable field ovens to bake bread. The sheet-iron Shiras ovens provided by the Subsistence Department (which supplied commissary needs to the U.S. Army) required pre-heating and skillful management. Soft bread was leavened with yeast, single-celled fungi (*Saccharomyces cerevisiae*) that in the absence of oxygen ferment starch and sugar and release bubbles of carbon dioxide. *Camp Fires and Camp Cooking* advised combining yeast strains used in brewing with an infusion of hops flowers (*Humulus lupulus*), which were long used in flavoring beer, a traditional method used in many homes. The antimicrobial phenolic compounds of hops curbed bacterial growth, which allowed the yeast to ferment in the dough without microbial competition. Bread dough was prepared from yeast combined with cooked potatoes to stimulate fermentation, and then the yeast "ferment" was mixed with a barrel and a half of flour and a few pounds of salt. The dough was left to rise ("prove") and then shaped into loaves. Three batches of dough yielded about 900 standard loaves in a day, enough to feed a regiment and much preferred to hardtack.[49]

Published by the U.S. Quartermaster General, *Bread and Bread Making* (1864) provided detailed information about the nutritional values of flours milled from regional crops of wheat, corn, and oats.[50] The authors concluded that southern wheat had the ideal proportions of gluten and starch for baking perfect loaves, ironic because cornbread was the preferred regional bread in the South. Union troops captured and consumed Confederate grain supplies, but whether they noticed a difference in the quality of the bread baked from southern wheat is unknown. Social acceptance was another factor even in times of high need. Until well after the Civil War, oats were regarded primarily as fodder, and many shunned the oat flour as an ingredient for bread baking. Fermentation remained a biological mystery because the metabolic role of yeast in converting sugars into alcohol in the absence of oxygen (the definition of fermentation) was poorly understood. The *Report of the Commissioner of Agriculture for the Year 1864* described fermentation trials that varied with the microbes cultured, growth conditions, sources of sugar and starch, and presence of inorganic matter such as metals and sulfur.[51]

Leavening was the major difference between soft bread and hardtack, which was more easily transported and stored for long periods. The G.H. Bent Company became one of the major suppliers of the standard square hardtack crackers of the Civil War era, and production increased in proportion to the number of soldiers who were on the march and unable to bake loaves of conventional bread. Known among soldiers as "teeth dullers," hardtack only hardened with age; the crackers were best soaked or crumbled in hot coffee or stew. Insect pests were common and included weevils and their larvae; a parody of Stephen Foster's "Hard Times, Come Again No More," the lyrics of "Hard Tack, Come Again No More" mention "old and very wormy" crackers. Men soon learned that larvae floated to the top when hardtack was immersed in liquid.

Cooking on their own, soldiers devised recipes to make hardtack more palatable. Softened in water and then fried in bacon fat, the crackers were known as "skillygalee" and notoriously indigestible. Puddings were made by combining powdered hardtack with fruit. Simmered in water or broth, hardtack dissolved and thickened gruel, soup, and stew. These one-pot meals were standard fare; *Camp Fires and Camp Cooking* provided methods for preparing soup and stews from meat, legumes, and other vegetables. "Beef Soup with Desiccated Mixed Vegetables" was a typical military meal that was prepared in large kettles for hundreds of men.[52] It made use of the dehydrated vegetables issued as part of the military ration; these comprised a mixture of dried carrots, cabbage, turnips, parsnips, and onions, baked as large blocks and then cut into cubes for distribution. Crops grown to maximum size become fibrous or woody, so the dehydrated vegetables required hours of simmering to soften. Soldiers referred to the "desecrated vegetables" in their stews, but the unappealing cubes provided vitamins A and C which helped to prevent night blindness and scurvy.

Legumes replaced meat in many military meals because their cotyledons (the enlarged leaves of the embryo) store high levels of both starch and protein. The seeds alone were consumed because the pods of early cultivated legume varieties were tough. When prepared properly with a bit of meat, dried beans became a palatable dish favored by many soldiers, but novice cooks often did not realize that they required careful sorting and washing to avoid stones and dirt. Dried beans needed overnight soaking in water prior to cooking, which may explain the note in *Camp Fires and Camp Cooking* that "beans badly boiled, kill more than bullets."[53] Without this step, which softened the starch and cell walls, beans often remained hard despite hours of cooking. Regional water differences also affected cookery, which troops may have noticed as they marched across states. Dried beans soaked in water with a high concentration of calcium (including water from limestone aquifers in Pennsylvania, West Virginia, Virginia, North Carolina, and South Carolina) were difficult to soften because calcium ions toughen seed coats. In these areas, army cooks needed more time to simmer beans until they were soft.

Recipes included beans for breakfast, bean soup, and baked beans, all prepared from varieties of *Phaseolus vulgaris*, the beans originally cultivated by Native

Americans. By the mid-nineteenth century, there were several cultivars available with both bush and climbing growth habits. *The Field and Garden Vegetables of America* (1865) listed over sixty "American garden beans" including yellow, cranberry, kidney, re-speckled, crimson, and black-eyed cultivars.[54] However, traditional baked bean recipes favored the small white climbing varieties known as pea beans. Army cooks baked beans in kettles positioned in fieldstone ovens or stone-lined holes covered with boards. The process required sorting, washing, and soaking the beans overnight, and then building a fire to heat the rocks hours prior to baking. Cooking often occurred overnight, so a meal of baked beans required two days to prepare. A popular song titled the "Army Bean" commemorated the small white beans common in Civil War rations:

> 'Tis the bean that we mean,
> And we'd eat as we ne'er ate before,
> The Army bean, nice and clean,
> We will stick to our bean evermore.

Soup made from dried peas (*Pisum sativum*) was recommended to army cooks but not well received by soldiers, although standard methods suggested flavoring the soup with pork. Dried peas had the seed coats removed and the cotyledons separated, and thus the split peas cooked rapidly into a porridge-like mush that often scorched in army kettles. *Camp Fires and Camp Cooking* noted that "for some unexplained reason, this article is by no means popular with the troops, and large quantities are constantly returned to the commissary as company savings."[55] Peas thrived in the early spring weather of both Union and Confederate states; supplies were abundant, but army cooks struggled to prepare peas over wood fires with uncontrollable heat. Peas had been a colonial favorite, and during the Civil War home cooks still used dried split peas to make into puddings which were boiled in cloths in the traditional English method.

As a stimulant, coffee (*Coffaea arabica*) fueled the war. Men drank it while on the march, at rest, before battles, after battles, and even while under fire. The high caffeine (1–1.5 percent) content remedied fatigue and encouraged endurance; boiling killed water-borne microbes, and the flavor masked the taste of muddy water. As part of their monthly rations, Union troops received a few pounds of coffee beans, which men roasted and then pulverized with rocks or rifle butts. Once ground, the coffee was boiled directly in water to make a strong brew. Typical coffee consumption among Union troops was about eight cups daily, and many considered it a nerve tonic without which they could not endure the hardships of war. Relief agencies such as the United States Christian Commission, a volunteer Protestant charity, provided comforts to soldiers during the Civil War, including coffee delivered to soldiers in the field. The Commission operated coffee wagons that were based on horse-drawn ammunition caissons, a design patented by John Dunton in March 1863. Each wagon carried several boilers, each with its own firebox and spigot, with an hourly coffee production of a hundred gallons.

The Union blockade prevented most coffee from being shipped to southern states; little was available for Confederate troops, and then it was used medicinally as a stimulant. There were occasional captures or trades in which soldiers exchanged tobacco in return for coffee, but most Confederate troops brewed coffee substitutes. These included a remarkable diversity of reasonably palatable plant parts, which were roasted until dark, pulverized thoroughly, and infused in water. The list of coffee substitutes included grains (corn, rice, wheat, and rye), roots (beets, chicory, dandelions, and sweet potatoes) and the nuts and seeds of various plants (acorns, peas, peanuts, sugar cane, okra, cotton, and asparagus). Certainly the most familiar coffee substitute was chicory (*Cichorium intybus*), a European composite (family Asteraceae) that became widely naturalized in North America; its taproots had been used for years to adulterate or stretch coffee supplies.

The roasted roots of chicory were a palatable coffee alternative when the Union blockade prevented coffee imports into southern states. Confederate troops relied on brewing chicory, a widely naturalized European composite that had been used for years to adulterate or stretch coffee supplies (Harter, 1998).

Folk wisdom suggested additional options, for instance seeds of the honeysuckle relative known commonly as wild coffee or feverwort (*Triosteum perfoliatum*). The pulverized seeds were infused in hot water into a palatable beverage, but the species had medicinal uses as well. Native Americans used decoctions of the plant to treat fevers and as a purgative. However, common names can deceive; Kentucky coffee trees (*Gymnocladus dioicus*) were not used as a coffee substitute during this time. Early settlers had experimented with grinding and brewing the seeds, but soon learned that the large seeds are toxic until roasted. Other options for brewing may have been more appealing, or perhaps the limited floodplain range of Kentucky coffee trees made the seeds difficult to obtain. The large leguminous fruits have a sweet pulpy layer, probably an adaptation to dispersal. Some have suggested that the trees are among the anachronistic species once dispersed by wooly mammoths, mastodons, and ground sloths, the extinct megafauna that disappeared by the end of the Pleistocene epoch about 13,000 years ago (see Chapter 2).

Francis Peyre Porcher encouraged coffee-starved Southerners to experiment with other possible substitutes, including the seeds of milk vetch (*Astragalus* spp.) and the related genera *Lathyrus* and *Vicia*.[56] He noted that *Astragalus boeticus* was used to make a coffee-like brew in England and Germany, but these suggestions were potentially risky. He was apparently unaware that consuming large quantities of *Lathyrus* can result in leg paralysis and other neurodegenerative symptoms, a condition now known as lathyrism and caused by the toxic amino acid ODAP (oxalyl diaminopropionic acid). *Astragalus* species were also potentially toxic because some concentrate selenium and others produce neurotoxins. Coffee cultivation was another option, although unlikely given the stringencies of wartime agriculture. However, some southern growers may have considered coffee as a possible crop for Confederate consumption; indeed Porcher recommended that planters consult the 1858 agricultural reports from the U.S. Patent Office, which documented practices used in coffee cultivation in South America and Jamaica.[57]

Whiskey, Wine, and Beer

Dr. Benjamin Rush predicted the evils of alcoholic drink in "A Moral and Physical Thermometer," published in 1790 as part of *An Enquiry into the Effects of Spirituous Liquors on the Human Body and Mind.* He charted spirits ranging from fermented cider to rum and attributed vice, debt, imprisonment, idleness, melancholy, murder, and disease to their consumption. Nevertheless Scottish and Irish settlers carried their knowledge of distilling whiskey from fermented grains to Southern colonies, where the temperance movement was not yet a major social influence. By the mid-nineteenth century, temperance societies described the perils of drink, but commercial distilleries and breweries operated widely. In 1851 the remote state of Maine was the first to legislate prohibition, perhaps because the high populations of sailors and loggers caused problems with public drunkenness.

Civil War soldiers endured hunger, cold, and other deprivations, and they drank whiskey for relief. Dram shops opened around army camps and sold whiskey to anyone who could pay. There was concern for soldiers' well-being and for the impact of alcohol on military readiness to fight; a note in the *Southern Cultivator* reminded readers, "Of all villainous concoctions, the liquors sold by camp followers are the most detestable and dangerous ... sure to be taken just when they should not be. Every soldier who means to do his duty to his country should insist that all vendors of these poisons be drummed out of camp."[58] Most southern whiskey was made from corn mash in which yeast fermented starch and sugar into alcohol. Distillation involved heating the fermented mash and collecting droplets of alcohol, which boiled at a lower temperature than water. Distilled alcohol was used to fortify concentrations of whiskey to higher alcohol levels than homemade wines and ales. The traditional method of aging in wooden barrels helped to develop characteristic flavors, but wartime whiskey was made in haste from whatever was available for fermentation and distillation, including wheat, peas, potatoes, molasses, and sugar.

Farmers soon realized that grains were worth more for distilling whiskey than for food or fodder, and concern mounted about staple crops. In North Carolina, the Confederate distillery drove corn prices so high that local civilians feared famine, but production continued to supply medicinal alcohol for army hospitals.[59] Virginia banned whiskey production from corn, wheat, and rye in 1862, but then revised the law to allow distillation of medicinal and chemical alcohol. The state of Alabama seized stills used in making whiskey, and the governor of Georgia halted all trains carrying whiskey and insisted that army regulations on drunkenness be enforced. The agricultural solution was to plant more corn for food and fodder and fewer cash crops such as cotton, a strategy that some growers resisted despite thousands of bales of stored cotton that could not be shipped due to the wartime Union blockade. Meanwhile in response to shortages and high prices, soldiers devised ways to ferment alcohol from rations or raided foods and sought whiskey from small-scale producers. Private stills, many in remote areas and valleys, produced wartime whiskey from grains and virtually any other plant material that yeast could ferment, including dried fruit, sweet potatoes, and pumpkins, alone or in various combinations.[60]

Cordials were considered medicinal for use in homes and hospitals, and they were a means to dilute distilled alcohol for moderate sickroom consumption. Francis Peyre Porcher described a "cordial for sickness in the army" in which sweetened blackberry juice was spiced with cinnamon, cloves, and nutmeg, boiled, and then fortified with whiskey or brandy made from distilled wine. Cherry cordial was made from an infusion of cherries in whiskey, which was then sweetened with brown sugar and boiled until thickened.[61]

Wine-making was part of the southern tradition. Colonists arriving in Virginia in 1607 planned to cultivate grapes, ferment the juice, and export bottled wine to England. Men were required to plant grape vines (twenty for each male over the age of twenty years), but fungal disease and insect attacks diminished the crops. Some soon realized that rather than planting European grapes (*Vitis vinifera*), southern growers should cultivate grapes native to North America. Thomas Jefferson hoped for successful Virginia vineyards to replace some of the dependence on tobacco crops. He sought the advice of European experts, including Phillip Mazzei of Tuscany who experimented with growing native North American grapes at Monticello. By the first half of the nineteenth century, Virginia was a successful wine-producing state known for abundant crops of grapes used in making claret, first developed and propagated by Dr. Daniel Norton of Richmond. His cultivar resulted from crosses among wild varieties of native summer grapes (*Vitis aestivalis*), a widely distributed species adapted to conditions in the eastern United States from Maine to Florida and as far west as Oklahoma. As a result of hybridization, 'Norton' vines easily tolerated the cold winters and hot, humid summers of the Virginia climate.

During the 1860s, nurseries advertised diverse grape varieties, including the Catawba varieties that were hybrids of European grapes (*Vitis vinifera)* and fox grapes (*Vitis labrusca*) native to eastern North America. Newer cultivars included 'Brinckle,' 'Union Village,' 'Cuyahoga,' 'Alvey,' and 'Ontario,' which were advertised for use as both table and wine grapes. Many have since disappeared, and their parentage is unknown; most were presumably hybrids. Nurseries typically supplied vegetative cuttings and rootstock, which preserved hybrid traits that might be lost if grapevines cross-pollinated and grew from seed. Southern growers were encouraged to plant grapes as part of the move to Confederate independence, and regional agricultural journals supplied information on grape culture and winemaking. Native muscadine grapes (*V. rotundifolia*) were used to make homemade wines, which sometimes were nothing more than a barrel of fermenting grapes. The *Southern Cultivator* provided detailed instructions for using Catawba grapes to make sparkling wines that were "very nearly equal Champagne" by tightly corking the bottles following fermentation to contain carbon dioxide bubbles.[62]

Many commercial wines were adulterated, so winemaking at home was a reliable way to repurpose spare fruit to supply common needs. Wine was widely regarded as medicinal, even by temperance reformers, and was considered essential in sickrooms and military hospitals. "Receipts for Food and Drink for the Sick," the chapter on sickroom cookery in *Miss Beecher's Domestic Receipt-Book*, recommended various concoctions of wine mixed with tapioca, gelatin, milk, eggs, or sago starch to nourish the sick or chronically ill.[63] Of course, some of the wine made in southern states was also intended for table use, as the blockade stopped shipments of imported wine. During the 1860s, southern farm journals published frequent articles on fermentation, aging, and bottling of wine, in which methods, successes, and failures were documented and shared. Prudent household management required a supply of homemade wines, and cooks learned how to prepare juice (known as the must) by crushing whole fruit. The outer waxy epidermis harbors wild yeast (*Saccharomyces ellipsoideus* and other species) that ferment fruit sugars into alcohol; however, wild yeast succumbed quickly to low levels of waste products from fermentation, so homemade wines contained just a few percent alcohol.

The characteristic deep hue of grape wines originates with anthocyanin pigments in red and purple grape skins, but other deeply pigmented fruit such as blackberries and raspberries (*Rubus* spp.) imparted deep color to homemade wines. Other fruits used for household wines included elderberries and elderberry flowers (*Sambucus* spp.), gooseberries and currants (*Ribes* spp.), cherries (*Prunus* spp.), tomatoes, apples, and the leaf stalks of rhubarb (*Rheum* sp.).[64] In the South, Union troops often destroyed or foraged fruit from grape arbors and orchards, requiring home cooks to seek alternatives for winemaking. During the early 1860s, *Godey's Lady's Book and Magazine* printed numerous methods for fermenting wine, including some that were based on using unlikely ingredients on hand. Parsnips required four days to ferment, followed by a year or more of aging to produce a palatable wine. Another option was walnut wine, which required no walnuts; it was made by steeping walnut leaves in water sweetened with sugar and honey, which yeast fermented into alcohol.[65] The leaves of native black walnut trees (*Juglans nigra*) were easily foraged, and their complex chemistry of terpenes and flavones provided flavor to homemade wine.

Various types of beer were also made at home, with no particular claims of medicinal value. Traditional brewing methods depended on fermenting malted barley (sprouted and dried grains of *Hordeum vulgare*) with hops (*Humulus lupulus*), which served as a preservative and flavoring. When barley supplies were low, wartime beer was fermented from corn, persimmons (*Diospyros virginiana*), the legumes (pods) of honey locust (*Gleditschia triacanthus*), and the starchy rhizomes of false chinaroot or China briar (*Smilax pseudochina*). Sassafras shoots (*Sassafras albidum*), leaves of Jerusalem artichokes (*Helianthus tuberosus*), and spruce essence (an extract from the needles of black spruce, *Abies nigra*) were used to flavor beers brewed at home and in army encampments.[66] Methods varied, depending on the available ingredients. Persimmon beer typically combined the fruit pulp with wheat bran and hops, but wartime brewers fermented a simple combination of persimmons and straw, perhaps with the addition of apple peelings, sweet potatoes, or honey locust pods.[67] Honey locust was easily foraged; their sugar-rich fruits likely evolved as adaptations to dispersal by the extinct megafauna that disappeared by the end of the Pleistocene epoch about 13,000 years ago (see Chapter 2).

African Crops and Diets

Civil War diets depended on several African plants that arrived in North America with the slave trade beginning in the seventeenth century. Imported crops included cowpeas (also known as black-eyed peas or crowder peas, *Vigna unguiculata*), a legume in which each pale seed has an eyespot surrounding the hilum, the attachment point to the pod wall. The seeds of some early varieties completely filled or crowded each pod, hence the origin of the name crowder pea. The center of cowpea genetic diversity is western Africa, which is probably where crops were first cultivated thousands of years ago. Natural variation in the species is apparent from the diverse cultivars that have been selected in southern gardens, including shrub and vine growth habits, small and large-seeded varieties, and eyespot pigmentation ranging from black and brown to red, pink, and green.

Enslaved people introduced cowpeas to plantations and cultivated them for their edible seeds, now known to contain about 26 percent protein and comparatively high levels of lysine and tryptophan, the amino acids that are comparatively deficient in corn. They provided only about half of the B complex vitamins in garden peas (*Pisum sativum*). Cowpeas were soon cultivated in southern states for both human and animal consumption and adopted as a valuable fodder for Confederate horses (see Chapter 3). As legumes, the plants release nitrogen into the soil and so improve the growth of crops planted nearby. They thrive in nutrient-poor soils and tolerate drought conditions, making them well adapted to some plantation conditions. Various legends linked cowpeas to good fortune during the Civil War and may explain the tradition of eating this dish on New Year's Day. According to one tale, General Sherman's troops mistook cowpeas for fodder and left them undisturbed when they plundered supplies during their march to the sea; other interpretations link cowpeas to the joyous emancipation from slavery. Cowpeas were often cooked with rice to prepare hopping john, which first appeared in *The Carolina Housewife* (1847) and specified the red-pigmented variety of cowpeas flavored with mint and bacon.[68]

Peanuts (*Arachis hypogaea*) are South American legumes that Spanish and Portuguese traders and slavers introduced to Africa, and they arrived in North America with the African slave trade. They were known as groundnuts because following pollination, the flower stems elongate and push the developing legumes into the soil.

Peanuts became a reliable southern crop used as a table food, oil seed, and fodder to fatten hogs. Virginia and Spanish varieties were distinguished for their traits; although Virginia peanuts were larger, farmers and gardeners often preferred the erect growth and higher oil content of Spanish varieties. Enslaved people and former slaves cultivated peanuts in their gardens (see Chapter 6), and in the early twentieth century, George Washington Carver investigated peanuts for their high oil content and for processing a milk substitute.

Eggplants arrived from Africa along with cowpeas, yams, watermelons and okra. This cluster of plants were grown, harvested, and prepared by the enslaved women who prepared meals, and each played a significant role in the evolution of southern food traditions. Virginians confused unfamiliar eggplants (*Solanum melongena*) with melons or squash, but rather they are relatives of tomatoes and related species in the nightshade family (Solanaceae). Some colonists suspected that white eggplants were toxic, information perhaps gleaned from Gerard's *Herball* (1633) in which pale eggplants were described as having a "mischievous quality" and leaves resembling henbane, a plant long associated with strong analgesic and possibly hallucinogenic properties, depending on dose (see Chapter 5). Eggplants were first domesticated in India, and Persians later introduced them to Africa where they became a familiar crop. Carried to America during the slave trade, eggplants produced prolifically in southern vegetable gardens. The fruit are large berries that vary in color from white to deep purple, but deeply pigmented eggplants with dense white flesh were preferred for stewing, frying, and stuffing.

Yams are the starchy tuberous stems of species of tropical vines in the genus *Dioscorea*, which were a starchy staple in African diets for thousands of years. Yams were carried as provisions on ships in the slave trade, and enslaved people planted and propagated various species as a staple food, including water yams (*D. alata*) which had been introduced from the Pacific region to Africa. The tubers can weigh several pounds and require boiling or baking to breakdown the toxic alkaloid dioscorine; many species also require peeling to remove the outer layer containing crystals of oxalic acid. The common name *yam* sometimes also refers to sweet potatoes (*Ipomoea batatas*), but the species are entirely distinct.

Watermelons (*Citrullus lanatus*) are native to tropical and subtropical Africa, and they were probably the most recent species in the gourd family (Cucurbitaceae) to be cultivated in American gardens. The fruit type is a pepo, a leathery berry with seeds embedded in soft tissue that in the case of watermelons is about 90 percent water. In their dry native habitats, the high water content of watermelons helped their seeds to germinate and root, and coincidentally the fruit appealed to parched humans. The coarse vines produced prolifically even in poor soil, and enslaved workers valued watermelons as a water source and sowed seeds in gardens and near cotton and corn fields. Cultivars varied in shape, size, flavor, and color of the flesh and rind; the bright pigmentation of the flesh arises from lycopene, a carotenoid pigment that functions as an antioxidant by absorbing charged oxygen atoms (free radicals) that potentially can damage cells. Vitamin chemistry was not understood at the time, but watermelons provided vitamins A and C and helped to prevent the deficiencies that resulted in scurvy and poor eyesight. The low fructose concentration (about 7 percent) provided flavor, and Confederate army cooks used watermelon juice when it was available to sweeten fruit pies.

The irony of watermelon culture cannot be ignored. The fruit alleviated misery among enslaved Africans in the New World, but after the Civil War its consumption spawned racial stereotypes. Following emancipation, watermelons symbolized independence because freed workers were able to cultivate and sell their own crops. In fact, racial stereotyping involving the fruit may have begun in 1869 with an image in *Frank Leslie's Illustrated Newspaper* of freedmen trading in watermelons, the possible starting point for drawings and blackface performances associating a fondness for watermelon consumption with laziness, untidiness, and a cheerful demeanor.

Citron melons originated in Africa as a genetically distinct variety of watermelon (*Citrullus lanatus* var. *citroides*) in which a thick rind surrounds tasteless or inedible white flesh. In American homes, these became known as preserving melons, and they were used to make a substitute for the sugar-coated peels of true citron, a citrus fruit (*Citrus medica*) used in preparing cakes and confections. Cookbooks provided detailed instructions for soaking the pared rind in brine; stewing it along with some leaves from the vine, presumably to enhance its green color; and then preserving the rind in sugar syrup. Mary Randolph described an antebellum luxury dish in which sugared citron was combined with cream for freezing.[69] Its use in baking was common and included wedding cakes and tea cakes; Catharine Beecher (1858) described a dark fruit cake made with a half-pound of preserved citron.[70] Crops were cultivated in some northern gardens during the war years although it certainly was not an essential food.

Gherkins (*Cucumis anguria*) are another African species that became a pleasant addition to American diets, arriving by way of Caribbean islands on slave ships travelling to North American colonies. Gherkins are closely related to garden cucumbers (*C. sativus*), but the sprawling vines produce small spiny fruit that were cultivated particularly for pickling. Recipes were regional in nature, but at minimum pickling required vinegar and salt, along with herbs and spices that provided flavoring and antimicrobial properties. Wartime conditions caused shortages in cloves, nutmegs, and other tropical imports, but temperate herbs such as mustard, bay laurel, and horseradish were available. By the early 1860s, *Godey's Lady's Book and Magazine* recommended a simple brine of only salt and strong (distilled) vinegar for pickling gherkins, probably an adaptation to wartime shortages of tropical spices.[71] In order to conserve the flavors of imported spices, cooks traditionally boiled down the leftover brine and used it as a sauce.[72]

Sesame (*Sesamum indicum*) is native to Africa, where local people valued its edible seeds that could be consumed much like a grain. Sesame plants arrived in Southern colonies with the African slave trade, and the seeds (known as benne) were soon baked into bread, boiled with greens, and simmered in broths. Sesame seeds slightly thicken cooking liquids, and the pressed seeds yielded an oil that replaced olive oil in cookery, an advantage after olive trees failed to grow in southern climates. Thomas Jefferson encouraged the use of sesame seeds and oil, and enslaved Africans used them in plantation kitchens, where benne became part of southern cookery. Sesame also had medicinal uses in preventing dehydration, and sesame water was valued as a dysentery cure (see Chapter 5). Okra (*Abelmoschus esculentus*) was also used as a thickening agent, particularly in Creole cookery, because the capsules release mucilaginous carbohydrates when cooked in water. The origin of okra may be African or Asian, and its seeds arrived in North America with the slave trade. The plants thrived in warm southern habitats, but when okra was not in season southern cooks used filé powder, the spice and thickening agent made from the dried leaves of native sassafras (*Sassafras albidum*), which early explorers mistook for cinnamon.

Diets varied with plantation economy and the ability of enslaved people to garden independently, but corn and pork were provided universally. Frederick Law Olmsted observed enslaved workers' diets in Virginia, where weekly allowances of cornmeal, bacon, and molasses provided most of the calories. He noted that in summer enslaved people could always grow as many vegetables as they wanted, but that many relied on diets of mostly cornmeal and bacon.[73] Owners who did not provide sufficient food for workers were held in low regard by their peers, but cornmeal allowances distributed to enslaved families could be traded for whiskey, which no doubt resulted in cases of deprivation. There were variations in preparing cornmeal; Olmsted noted that enslaved people in Georgia and South Carolina preferred cracking and grinding dried corn into their own meal using millstones, which they believed baked into a sweeter bread than conventional cornmeal.[74] Rice was also part of the food allowance for enslaved people who worked in paddies. Wheat was a rarity reserved for Christmas and other celebrations, although some bread was made from flour purloined from plantation storerooms.

Cooking was done in fireplaces, either centrally or in individual cabins, depending on circumstances and preference of the plantation owner. Firsthand information about diets can be gleaned from the narratives transcribed by the Federal Writers Project (1936–1938), part of the Works Project Administration.[75] Meals were served in bowls or communal troughs made of gumwood (*Liquidambar styraciflua*), a hardwood tree with a wide southern distribution. Bowls, cups, and ladles were crafted from bottle gourds (*Lagenaria siceraria*).[76] Gourds were also used as *minkisi* (*nkisi*), containers that held items associated with spiritual and healing practices; these were often botanical in nature and included charcoal, plant parts, and nuts.[77]

Recollections from former enslaved people reveal the botanical foods and degree of self-sufficiency that they had in devising meals. Ash cakes were made from a batter of cornmeal, water, and lard baked on poplar and chestnut leaves. Typical cooking methods centered on long-simmered soups and stews of vegetables and salted pork or bacon. One-pot meals were typical; a familiar stew was known as *sofki*, which was prepared with ground corn and hickory nuts boiled in water, enriched with the addition of greens and other available ingredients. Enslaved women often used pot-scrapings and leftovers from the kitchen to prepare one-pot meals known as *juba*.[78] People enslaved to work in houses often had access to more and better food than field workers.

Enslaved people with the most adequate diets had access to fresh crops from shared or individual garden plots, but some may have preferred to sell rather than eat the crops that they grew. On some plantations a gardener, often an older field worker, tended a central plot that yielded a shared harvest. Alternately families cultivated their own smaller plots of cowpeas, beans, sweet potatoes, turnips, pumpkins, and various greens, the diverse harvest known generally as "garden sass." All of these crops had high yields and were comparatively easy to cultivate. Onions, okra, and squash were also planted, but less frequently. Corn was grown in large fields; enslaved people participated in the labor of producing their corn allowance, and they observed corn-shucking events as celebrations. Shucks were saved for weaving floor mats and horse collars, crafts that were taught to young children.[79] Some former enslaved people also mentioned cultivating small plots of sugar cane, and garden hives provided honey (the concentrated nectar of local bee-pollinated plants), but this may have been uncommon.

Cultivated greens were typically collards, derived from *colewort*, an old term for cabbages and related vegetables that originated in Europe. Garden greens included leafy varieties of *Brassica oleracea* (the widely cultivated species that includes kale, cabbage, and related varieties) or the shoots of turnips (*Brassica rapa*). The thick leaves of collards were typically cooked and flavored with pork, and they were a valuable seasonal source of vitamins A, C, and K, now known to be essential in blood clotting. Other greens such as wild mustards may have been foraged, as were blackberries, hickory nuts, and walnuts, depending upon access to forests and fields.

Sweet potatoes (*Ipomoea batatas*) were traditionally baked in coals and provided essential carbohydrates and vitamins in the diets of enslaved people. The species is related to morning glories (family Convolvulaceae) and likely originated in Central or South America; they were probably first grown in North America after pre–Columbian times. Virginia colonists cultivated sweet potatoes in the early seventeenth century, and New England farmers had successful crops as early as 1764. The plants were propagated asexually from cuttings that rooted readily and grew prolific vines. Early sweet potato varieties had red, purple, orange, and white storage roots, with vitamin content that varied with pigmentation. The roots were an important source of vitamin A, but pale varieties with fewer carotenes (the molecular precursor to vitamin A) were often preferred and cultivated by enslaved workers. Indeed some dark-pigmented varieties of sweet potatoes may have provided forty times more vitamin A than what was found in paler cultivars.[80] A similar situation existed with corn; some enslaved people preferred white grains and cornmeal that coincidentally provided less vitamin A than the yellow varieties cultivated for food and fodder.

Protein deficiencies may have occurred, depending on the availability of legumes as an alternate protein source when meat was meager or fatty. Cowpeas and other legumes were dried, and some fresh crops were stored, depending on local practices. As the primary source of vitamin A, sweet potato access year-round was also essential, but it is unknown how many enslaved people had consistent access to such basic nutrition. Sweet potatoes were sometimes stored outdoors in straw-lined pits or embanked in soil, sand, and leaves.[81] Pit cellars excavated in cabin floors were used to keep sweet potatoes, turnips, pumpkins, cabbages, and apples, which troops found and raided during the war years.[82] Despite their importance as root cellars, some planters objected to excavation inside cabins and responded by raising cabins on piers, perhaps because interior pits provided hiding places for stolen goods.[83]

Stored crops provided the only botanical foods during the late fall, winter, and early spring, so some (and possibly most) enslaved people survived largely on a diet of cornmeal and fatty pork for half of the year. Potato peelings and roasted corn were brewed as coffee substitutes.[84] Anemia may have resulted when iron-rich molasses was replaced by sorghum syrup, made from a grass crop with African origins (*Sorghum bicolor*, see Chapter 3).

Plantation owners controlled food access, and enslaved people may have suffered from pellagra, beriberi, and scurvy as a result of dietary deprivations. Generally speaking, diets depended on botanical foods, primarily corn and sweet potatoes that are rich in carbohydrates and low in protein. There is also speculation that long cooking times depleted vitamin contents and that rancid pork fats caused the depletion of vitamin A.[85] There were seasonal variations in the available botanical foods, with vitamin deficiencies associated with scurvy and night blindness more likely when fresh crops were unavailable. Some plantation owners provided orchard fruits such as apples, pears, peaches, and cherries, while others maintained that fruit caused illness among enslaved workers and forbid its distribution. Diets highly dependent on corn and cowpeas were often lacking in B complex vitamins. Thiamine (vitamin B1) was needed to prevent beriberi, which potentially had severe effects on the nervous and cardiovascular symptoms. Niacin (vitamin B3) deficiency caused pellagra, with symptoms that included dysentery, dermatitis, and dementia. The condition was poorly understood and may have been confused with malaria, diphtheria, typhus, or typhoid fever.[86] It is possible that geophagy, a form of pica in which soil is consumed, resulted from nutritional cravings caused by various dietary deficiencies.

Survival

The Civil War food supply was often unreliable, particularly in Confederate states. Hunger loomed as Southern agriculture dwindled and Union troops plundered farms and stored foods. Past practices and experimentation were essential to survival, including much of the wisdom that Frances Peyre Porcher compiled in *Resources of the Southern Fields and Forests* (1863). Sorghum seeds, wild rice (*Zizania* spp.), and acorns (*Quercus* spp.) were known emergency foods that were ground into flour when supplies were short. Depending on the oak species and seasonal variations, acorns varied in their tannin content; some may have been too astringent to consume, which explains the recommended addition of barley flour to acorns milled for bread-baking. Porcher also suggested milling various starchy rhizomes into flour, but several of the species that he listed were potentially toxic, including autumn crocus (*Colchicum autumnale*), hellebore (*Hellebore* spp.), and mandrake (*Mandragora officinarum*). As a physician, Porcher knew these plants for their medicinal properties administered in controlled doses, and he advised soaking "the pulp in a considerable quantity of water" which might have removed some toxic alkaloids and glycosides.[87]

A safer option involved baking bread from wheat flour combined with sweet potato pulp.[88] Perhaps ironically, Porcher recommended stretching cornmeal supplies with rye flour to approximate Boston brown bread, which he described as "a useful hygienic preparation" made from two parts cornmeal and one part rye flour.[89] However, it is unknown how much rye grain was available in Confederate states and whether cooks were willing to modify traditional cornbread into distinctly Yankee loaves. In times of desperation, even corn waste had uses; the shucks were used to make brooms and processed with lye and fat into soap, the cobs were used as fodder, and the ashes of corn cobs were used to produce chemical leavening, a homemade baking powder similar to the potash from wood ashes.[90]

As apple orchards were destroyed, cooks made vinegar from other sugar-rich fruit and plant parts, including figs, beets, grapes, persimmon, sugar cane, watermelon, sumac (*Rhus glabra*), and mayapples (*Podophyllum peltatum*). Alternately, a mixture of molasses, water, and yeast was allowed to ferment in a warm place until the sugar solution converted to vinegar through the action of wild yeast and acetic acid bacteria.[91] Even edible oils were in short supply; growers in South Carolina struggled with cultivating olive trees (*Olea europaea*), which grew very slowly in the southern climate. Oil pressed from poppy, mustard, and sesame seeds eventually substituted for imported olive oil.

A similar situation occurred with imported tea. Colonial attempts with growing Asian tea (*Camellia sinensis*) had failed due to climate differences, but in 1858 the U.S. Department of Agriculture arranged for British explorer Robert Fortune to provide tea seeds collected in China. These were grown into saplings for tea plantations across the Southeast, but the project again failed; those established near Knoxville were a short-lived experiment during the 1860s.[92] War time notes on tea culture from the federal Department of Agriculture reflected contempt for southern agricultural practice. The *Report of the Commissioner of Agriculture for the Year 1863* noted that the trees grew well in the South and that failure of southern tea crops was caused by nothing more than neglect, attributed to "the peculiar social condition of the southern population, semi-barbarous, with but few educated dominant minds, averse to improvement or to change of any kind."[93]

The Union blockade prevented tea imports from reaching Confederate states, so familiar kitchen herbs were brewed depending on availability and personal preference. Some wild plants were also harvested as tea substitutes, including blackberry leaves, sumac berries, and the southeastern hollies known as yaupon (*Ilex vomitoria* and *I. cassine*). Sassafras roots, long brewed by enslaved people as a blood tonic, were another option; the distinct flavor of sassafras tea results from volatile oils that contain terpenes, but it is now considered potentially carcinogenic based on safrole content in the oil produced by sassafras bark.

Of course, none of the alternative tea plants provided the same chemistry as true Asian tea. Wartime notes from the U.S. Department of Agriculture spoke disparagingly of the common shrubs known as New Jersey tea (*Ceanothus americanus*), which apparently some mistook for true Asian tea. The unrelated native shrubs lacked "the principle of Theine, to which the virtues of tea are ascribed," yet the tea-like leaves did contain tannins and several alkaloids (ceanothine, pandamine, and others) with mildly stimulating effects.[94]

Agricultural journals suggested experimentation with new crops, particularly weedy species that thrived with minimal cultivation. One option was Jew's mallow (*Corchorus olitorius*), a Syrian import grown as a potherb to stew with meat. Southern cooks soon realized its similarity to okra (both are species in the mallow family, Malvaceae) and so simmered the young capsules in soups.[95] Ground cherries (*Physalis* spp.) became popular as a fruit for pies, and they replaced true cherries as Union troops plundered and destroyed orchards. An article in the *Southern Cultivator* claimed that a single sprawling vine produced quarts of fruit, described as "one of the most delicious of fruits" that could be stored fresh for months.[96] Their unique appearance also attracted attention; each fruit resembled a small, yellow tomato surrounded by an inflated husk. Like tomatoes and Irish potatoes (all are species in the nightshade family, Solanaceae), ground cherries required minimum cultivation and thrived in southern heat. However, for many civilians and freedmen, the cultivation of any food crops became impossible. With homes destroyed and fields and farms plundered, many survived by foraging for wild or naturalized food plants, including hickory nuts, mulberries, blackberries, and the shoots and rhizomes of sugar cane that escaped from cultivated fields.

Southern agricultural journals suggested that farmers cultivate new wartime crops, particularly species that thrived with minimal cultivation. Some experimented with Jew's mallow, a Syrian weed grown as a potherb; cooks soon realized its similarity to okra and simmered its capsules in soups (author's collection).

5

Medicinal Botany and Medical Practice

Antebellum Practices

Nineteenth-century medical care was remarkably diverse, depending on location and philosophy for the type of medicine practiced. Rural families cultivated kitchen gardens that included both European and American medicinal plants, and they depended on many traditional homes cures used by early settlers. In contrast, by the 1860s, most families who lived in cities and towns purchased pharmaceutical drugs, most of which contained one or more botanical ingredients. Commercial pills and tonics were commonplace, and typical advertisements encouraged self-diagnosis and promised successful outcomes. Such products are now often known as proprietary or patent medicines (the bottle designs were patented) and almost invariably included medicinal plants or botanical extracts. Although their proprietary nature precluded full knowledge of the ingredients, profits from sales to trusting customers were enormous. In the North, revenue stamps on patent medicines supported the Union army, a tax that was eventually repealed in 1883.

Traditional cultivated medicinal herbs overlapped with kitchen herbs, including several European species in the mint family such as rosemary (*Rosmarinus officinalis*), thyme (*Thymus vulgaris*), peppermint (*Mentha piperita*), and sage (*Salvia officinalis*). They were widely used in brewing medicinal teas to sooth sore throats and improve digestion. Clary (*S. clarea*), closely related to sage, was used to treat various eye ailments, and tea steeped from horehound (*Marrubium vulgare*) was used to cure chills. Based on the potent terpenes in their essential oils, these various mints had antibacterial properties, which controlled microbial growth in preserved foods in the centuries before refrigeration and canning technology. As practical botanists, settlers recognized American mint species that could be added to the household pharmacopeia of known medicinal plants. For instance, bee balm (*Monarda didyma*) had known antiseptic properties (the result of the antimicrobial compound thymol), and it was widely adopted in the eastern U.S. used for treating sore throats and skin infections. However, potency and potential toxicity characterized many of the species that were most effective. The New World mint known as pennyroyal (*Hedeoma pulegioides*) resembled European pennyroyal (*Mentha pulegium*); both contained the terpene pulegone and were cultivated in kitchen gardens to use as both tea herbs and abortifacients.

Kitchen gardens often included Eurasian opium poppies (*Papaver somniferum*) used in preparing soothing syrups for teething children, but household manuals warned against overdosing because homemade poppy infusions were of uncertain strength. Cultivation was common enough that the minute seeds dispersed widely and poppies soon grew wild in local habitats. In the first edition of *A Manual of Botany of the Northern United States* (1848), Asa Gray noted that opium poppies had "escaped from gardens."[1] By the Civil War, he recorded naturalized populations growing near homesteads.[2] During the Civil War, opium poppies became the medicinal plant in greatest demand for military use.

Native American practices suggested additional plants for cultivation and adoption, including pinkroot (*Spigelia marilandica*) which was used to treat parasitic worms. Scottish botanist Alexander Garden (1730–1791) investigated its anthelminthic properties and published "An Account of the Indian Pink" in 1771. Pinkroot soon became a popular herbal remedy, but the alkaloid spigeline is potentially lethal, and mothers were cautioned against inadvertent overdoses when treating children for parasites. Native Americans used pokeweed (*Phytolacca americana*) as a salve to treat rheumatic pains, and nineteenth-century medical texts described its external use for pain and skin eruptions and internal use as an emetic and purgative. Mitchell's *Materia Medica and Therapeutics* (1850) described pokeweed as "most undesirably prevalent almost everywhere" for its tendency to self-sow on abandoned or disturbed land and along roadsides.[3] There was no need to cultivate pokeweed, and it was widely used because populations were nearly ubiquitous. Medical claims included curing erysipelas (a skin infection caused by *Streptococcus pyogenes*) and tetters (an antiquated term for several contagious skin infections, including herpes and ringworm) and implied possible antibiotic and antiviral activity.[4]

Those who consulted physicians encountered a diversity of medical perspectives. The heroic practices of the early 1800s were still in use during the 1850s and 1860s, including bloodletting, purging, and sweating to treat perceived humoral imbalances. Heroic practitioners administered strong laxatives such as castor oil, the oil pressed from seeds of castor bean plants (*Ricinus communis*). When supplies of imported castor beans ran short, gardeners in both the North and South grew the robust plants and harvested the medicinal oil. Southern growers lacked the hydraulic presses for large scale extraction, but agricultural advice columns suggested using a mortar and pestle to crush the seeds, following by boiling in water to separate the castor oil.[5] Known also as *palma cristi*, cultivated castor bean plants became familiar annual ornamentals in nineteenth-century gardens; some varieties were eight feet tall with red or purple-tinged leaves.

Another purgative was the emetic extract from mayapple (*Podophyllum peltatum*), a wildflower native to a wide range across the eastern U.S. Its rhizome slightly resembles the branched roots of the Eurasian mandrake (*Mandragora officinarum*), used since ancient times as an analgesic and panacea. As a result of this coincidence, mayapple became known as American mandrake, although the species are unrelated. Native Americans used it to treat warts and possible cancers, which is noteworthy in light of the modern chemotherapeutic drug etoposide that is made from its resin. Nineteenth-century practitioners of heroic medicine valued mayapple as an emetic and cathartic herb, and it was also used to treat jaundice, fevers, and sexually-transmitted bacterial infections that were widespread in wartime. Estimates suggest that 8 percent of Union troops were infected with venereal diseases, and other treatments were limited to a few botanical medicines (pokeweed, elderberry, and pine resin) and mineral medicines, which included mercury compounds and silver nitrate.

Practitioners of heroic medicine used mayapple extract for its emetic and cathartic properties, and it was one of several treatments for venereal diseases in Civil War troops. Known as American mandrake, its branched rhizome was thought to resemble the taproot of the unrelated Eurasian mandrake (Harter, 1998).

Samuel Thomson (1769–1843), a New Hampshire farmer who became an itinerant healer, adopted medical methods that he learned from local practitioners of herbal medicine. After experimenting successfully on himself and family members, he developed treatments that relied primarily on medicinal plants native to North America. In particular, Thomson's botanical cures provided an alternative to mineral drugs prescribed widely by heroic practitioners, including tartar emetic (antimony potassium tartrate, a salt of tartaric acid from fermented grapes) and calomel (mercurous chloride). He lectured and proselytized widely throughout New England, retailed the rights to practice his regimen of cures, and encouraged the formation of Friendly Botanic Societies to spread his populist medical mission further afield.

Thomsonian medicine depended on several plants with purported medicinal properties. Foremost was Indian tobacco (*Lobelia inflata*), a North American wildflower that synthesizes over fifty alkaloids in its roots. Lobeline in particular can slow heart rate, increase blood pressure, and cause seizures, all potentially deadly effects, but Thomson prescribed a tincture of Indian tobacco as a treatment for asthma. Long before he adopted it as one of his cures, Native Americans also used the species to treat respiratory ailments. Other Thomsonian cures were made from cayenne pepper (*Capsicum annuum*, native to southern regions of North America), goldenseal (*Hydrastis canadensis*, native to the eastern U.S.), and bayberry (*Myrica gale*, native to both northern North America and Europe). Later practitioners prescribed proprietary drugs that they advertised as "Thomsonian," and these included various wine bitters, rheumatic tinctures, ointments, liniments, alterative syrups, and nerve pills, all presumably compounded from botanical ingredients.[6]

Eclectic medical practice erred on the side of formal training in anatomy and physiology, and physicians were encouraged to select the most suitable treatments to restore health to their patients. Eclectic practitioners leaned toward botanical drugs and rejected the heroic practice of bloodletting and the overuse of calomel as barbaric; hospitals and institutes promoting an eclectic approach were established in New York and Ohio. Eclectic practice was first promoted in the early 1800s by Constantine Samuel Rafinesque, a botanist who studied the medicinal plants used by Native Americans. His followers pursued various botanical and Native American cures until the 1930s, when eclectic training was absorbed into mainstream medical practice. In fact, many of the treatments common during the Civil War years were condemned in the 1910 Flexner Report, an overview of medical training.[7] Such practices included homeopathy, eclectic medicine, and herbalism and centered on concern for the widespread use of unproven and untested botanical drugs.

Yet during the mid-nineteenth-century physicians and folk practitioners alike depended on medicinal plants for their most reliable drugs, including several imported species that became increasingly difficult to obtain during the Civil War. Ignoring the Confederate secession, Commissioner of Agriculture Isaac Newton encouraged cultivating a federal supply of well-known medicinal plants such as opium, castor bean, camphor, and asafetida in southern states, which he believed would provide ideal growing conditions for warm weather crops to grow in abundance. Imported opium was often of dubious quality and variable potency; as observed in the *Report of the Commissioner of Agriculture for the Year 1862*, "It is notorious that no drug is so generally adulterated as opium when received from abroad."[8] From the Union perspective, useful medicinal plants, foods, fibers, and other botanical commodities could potentially replace cash crops and enslaved labor in Confederate states.[9] In fact southern newspapers encouraged civilians to cultivate opium and castor bean—however, to supply the Confederate army and not to replace traditional cash crops after the war ended.

Southern Strategies

During the war years, medical practice in Confederate states relied on whatever healing expertise, pharmaceutical drugs, and medicinal herbs were available. At the time, most pharmaceutical products were European imports supplied to drug wholesalers located in the North. These firms employed pharmacists who prepared galenicals, the various tinctures and extracts used in compounding pharmaceutical medicines. Adulteration of highly valued medicinal plants was a frequent problem, with drug suppliers often making unspecified substitutions or stretching volumes with cheaper ingredients. At the peril of patients, the dried leaves of belladonna (*Atropa belladonna*) were often stretched with foxglove (*Digitalis purpurea*), which affected cardiac muscle. Pebbles and bullets increased the weight of opium packets, and various inert ingredients such as ground minerals increased both mass and volume of various drugs.

The Union blockade of shipments impeded the import of pharmaceutical drugs into southern states, resulting in severe shortages and the need for strategic alternatives. There were attempts to circumvent the blockade by sending Confederate agents to Europe, where they purchased dried medicinal plants directly rather than depending on shipments arriving from northern ports. Smugglers also soon realized that small quantities were easy to transport and turned high profits by bringing in essential botanical pharmaceuticals. Some European ships landed at southern ports, but more often they unloaded medicinal plants at the Confederate Medical Bureau in Bermuda, where southern botanical products such as turpentine, tobacco, and cotton were traded for critical drugs.[10] Opium and quinine were among the drugs in highest demand, but supplies that arrived despite the blockade were exorbitantly expensive or already destined for military use.

Native and naturalized wild medicinal species and cultivated herbs were the only alternatives for stocking the early Confederate pharmacopoeia. Physicians and pharmacists used any available medicinal plants to compound drug mixtures, often employing the same species used in local herbal practice. A common folk belief held that local flora was best for treating ailments and conditions typical of a region, so attention favored indigenous and naturalized plants for preparing various home cures. Folk wisdom also included remnants of the medieval Doctrine of Signatures, the notion that plant traits such as shape, taste, texture, or color may reveal medicinal uses. Available native plants such as blackberry, pokeweed, tulip poplar, jimsonweed, mayapple, elm, and sassafras were ingredients in various teas, tonics, syrups, salves, poultices, washes, and other homemade remedies. Botanical products including turpentine, vinegar, charcoal, sugar, and flour were also purported to have some curative properties; the *Confederate Receipt Book* described a diphtheria treatment made from turpentine and charcoal and vinegar-based cures for coughs and dysentery.[11]

Traditional garden herbs grew abundantly in the southern climate, providing a ready supply of fresh and dried herbs as the *materia medica* for physicians, folk healers, and women with families needing care. Even for city dwellers, some wartime medicine relied on familiar herbs with European origins, many cultivated in America since colonial times. Most herbs tend toward weedy growth, and they frequently escaped garden cultivation; naturalized wild populations were often harvested when gardens were neglected or destroyed. During the Civil War, household herbs comprised much of the southern pharmacopeia for ailments that included infectious diseases, fevers, melancholia, congestive heart failure, and wounds. Most traditional herbs contain several bioactive compounds, and some have some antibiotic properties. Virtually no herbs were cultivated or collected for just a single use, and wartime herbal medicine arose from a combination of plant availability and the need to improvise in difficult times. Herbs in the mint family (Lamiaceae) included thyme (*Thymus vulgaris*), peppermint (*Mentha piperita*), and lemon balm (*Melissa officinalis*), and all were widely used in teas and tonics for flatulence, stomach complaints, poor digestion, parasitic worms, headaches, and colds. In a similar vein, cultivated and naturalized onions (*Allium cepa*) and wild onions (*A. canadense* and related indigenous species) had several uses in folk medicine as cathartics, expectorants, and vermifuges. The *Confederate Receipt Book* mentioned using onion poultices to cure infections, revealing practical knowledge of the sulfur-containing compounds alliin and allicin that act against several bacteria and fungi.[12] When onion tissue is damaged, a series of chemical reactions converts alliin into allicin, which is antibiotic to both *Salmonella* and *Staphylococcus*.

No doubt there were varying degrees of efficacy, and some cures involving European herbs may have been nothing more than placebos. One such herb was lungwort (*Pulmonaria officinalis*), interpreted by the Doctrine of Signatures as having leaves that resembled lung tissue. Traditional uses of lungwort included consumption (tuberculosis) and other lung ailments, for which it is entirely ineffective. Saint John's wort (*Hypericum perforatum*) was traditionally used to treat wounds (suggested by the glandular dots and reddish oils in its leaves) with unknown efficacy. By the time of the Civil War, the herb was used to sedate and also to treat stomach complaints, effects presumably discovered through trial and error as a home cure. Valerian (*Valeriana officinalis*) was also widely used as a mild sedative, and some naturalized plants grew wild along roadsides in Union states. The root extract contained a complex mixture of alkaloids and terpenoids known as valepotriates and was used in treating various nervous complaints; valerian was adopted for military use in treating epilepsy, hysteria, spasms, and convulsions. The closely related species known as large flower valerian (*V. pauciflora*) was indigenous to Tennessee, but it is unknown whether it was ever adopted as an herbal replacement for increasingly scarce supplies of European valerian.

Confederate Medicinal Flora

The Confederacy ultimately resorted almost exclusively to indigenous and naturalized flora as the primary source of medicinal plants and drugs for military use. Despite a dire shortage of equipment, the Medical Department established laboratories in several cities, including Richmond, Charlotte, Columbia, Mobile, Macon, Atlanta, Arkadelphia, and Knoxville, tasked with investigating indigenous flora. The Surgeon General's Office in Richmond, Virginia, issued a pamphlet with descriptions of medicinal plants, doses, and properties, along with instructions for drying plants collected from wild sites.[13] Newspaper advertisements specified the medicinal plants needed by the Confederacy and the prices offered; a list published in the *Arkansas Gazette* (October 25, 1862) sought clean, dried collections of opium poppy capsules, dogwood bark, seeds and leaves of jimsonweed, henbane, pinkroot, boneset, and 60 other species, most with currently recognized medicinal value. The two volumes of the short-lived *Confederate States Surgical and Medical Journal* also published articles and communications on the use of indigenous medicinal plants for treating wartime ailments. The office of the Confederate surgeon general encouraged southern physicians to return to traditional remedies including bloodroot, lobelia, butterfly weed, pinkroot, and mayapple, which were all known and used by

Native Americans.[14] A root decoction of pokeweed was recommended as an external treatment for camp itch, a chronic skin condition that vexed many soldiers, and a tincture of the berries was suggested for its laxative effects.[15]

Francis Peyer Porcher, a South Carolina surgeon with considerable botanical expertise, used multiple sources to compile *Resources of the Southern Fields and Forests* (1863), and he included numerous likely substitutes and possible replacements for well-known medicinal species.[16] As an owner of enslaved people, Porcher had personal interest in the matter of Confederate victory and in contributing to the war effort and military medicine. He was excused from hospital duty as a surgeon to spend his time in compiling the only *materia medica* available to the Confederate military. Much of the information that he provided was gleaned from diverse sources with varying degrees of credibility, from classical botanical works and standard medical and chemistry texts to encyclopedias, patent office reports, dispensatories, medical and farm journals, gardening texts, newspapers, folk wisdom, and hearsay. Many of his suggested plant uses were speculations based on botanical relationships and evolutionary similarities, but whether Porcher accepted evolutionary theory is unknown. Charles Darwin published *On the Origin of Species by Means of Natural Selection* in 1859, just a few years before Porcher organized the southern flora by family.

Porcher's work supported the scheme developed by Confederate Surgeon General Samuel Moore to have physicians collect and ship medicinal plants to pharmaceutical firms for use in compounding drugs. However, not all North American species were well documented medicinally; doses and outcomes were far from definite, and species new to medicine required evaluation. The 1845 *Dispensatory of the United States of America* included an appendix of "Drugs and Medicines Not Officinal," which documented and described remedies not typically used in the pharmaceutical trade.[17] The list included numerous indigenous plants that were better known to folk healers than to practicing physicians. Yet faced with shortages of standard botanical remedies, the Confederate government sought many of these plants, including species now known to contain some of the most bioactive compounds.

Directions for collecting and drying plants were an essential part of Porcher's text. He realized the possible effect of timing on medicinal properties and recommended digging roots before annual plants flowered, harvesting biennials in their second year of growth, and collecting bark in the fall. He provided detailed instructions for thoroughly drying plant collections made for shipment to the Surgeon General's office, noting that "a carefully pressed specimen of the stem, leaf, and flower of each medicinal substance collected … should be obtained and forwarded with each collection, for the purpose of aiding in its identification."[18] These were all prudent practices; specimens confirmed identity when amateurs might easily collect the incorrect species, and microbial activity can dangerously alter the chemical constituents of plant tissue that is damp and moldy.

Opium, the botanical drug in greatest demand, was in chronic short supply although opium poppies were cultivated in southern gardens and naturalized in abandoned sites. There was no comparable alternative to morphine and other opium alkaloids, so pharmacists and practitioners sought other plants with analgesic properties. Porcher suggested using the leaves of water hemlock (*Cicuta maculata*), a relative of the hemlock (*Conium maculatum*) known to ancient Greeks and Romans as a suicide plant. Both are highly toxic, although the toxins differ. Water hemlock synthesizes cicutoxin, which affects the central nervous system and paralyzes respiration; European hemlock contains several toxic alkaloids (coniine and others) with effects similar to cicutoxin despite involving different receptors. Porcher described the effects of water hemlock as "powerfully narcotic, sedative, and anodyne" and stronger than opium in treating pain associated with cancers. However, safe doses were minute (one or two grains of dried leaf tissue, each equal to 0.0648 grams), so deadly overdoses were a definite concern.[19]

The two hemlock species may in fact have been confused and used interchangeably. The *Standard Supply Table of the Indigenous Remedies for Field Service and the Sick in General Hospitals* listed *Conium maculatum* as a "narcotic and sedative" despite its extreme toxicity, although medical texts also mention its use as a purported cure for cancer and erysipelas.[20] Porcher also suggested the native red buckeye (*Aesculus pavia*) as a possible narcotic; the seed extract can cause stupefaction (the effect of a high dose of the glycoside aesculin), but it lacked true analgesic properties. It was also used externally to treat gangrenous ulcers.[21]

Another opium substitute was derived from the resin of European hops (*Humulus lupulus*), which is now recognized as having sedative, soporific, and antimicrobial qualities. The climbing plants had long been cultivated for beer-making and were easily collected from naturalized populations across Confederate states. Porcher mentioned infusions of hops as a substitute for laudanum (a tincture of about 10 percent opium dissolved in ethanol) and provided detailed instructions for its cultivation.[22] It was listed as a hypnotic in the *Standard Supply Table of the Indigenous Remedies for Field Service and the Sick in General Hospitals*.[23] The *Dispensatory of the United States of America* (1845) described garden lettuce (*Lactuca sativa*) as having qualities similar to opium.[24] The dried juice resembled the brownish shade of true opium and was sometimes known as lettuce-opium with "the anodyne and calming properties of opium, without its disposition to excite the circulation, to produce headache and obstinate constipation."[25] Lettuce may indeed have provided mild

Water hemlock was regarded as a possible substitute when opium was in short supply; the plants produce a potent toxin that affects the central nervous system. Grains of dried leaf tissue were administered with care, and deadly overdoses were possible (Harter, 1998).

sedation due to an alkaloid similar to hyoscyamine (typically found in several potent medicinal plants, including henbane and mandrake), but it lacked the powerful analgesic effects needed to treat extreme pain.

Aside from opium and purified morphine, the most effective analgesic drugs relied on nightshade species (family Solanaceae) known for their complex tropane alkaloid chemistry. Henbane (*Hyoscyamus niger*) did not naturalize, but limited supplies were available from cultivated gardens or blockade runners, the Confederate agents who obtained and shipped European drug plants. Jimsonweed (often known medicinally as "stramonium," *Datura stramonium*) was a North American alternative to henbane; its vernacular name recalled an incident at Jamestown, Virginia, in 1705, when British soldiers ingested the spiny fruit and hallucinated for several days. Controlled doses of hyoscyamine, atropine, and other tropane alkaloids were used medicinally to relieve pain (or perhaps quell the memory of pain); uncontrolled doses were powerfully hallucinogenic and potentially deadly. Porcher neglected to mention nightshades as possible opium substitutes, but both henbane and jimsonweed were listed in the *Standard Supply Table of the Indigenous Remedies for Field Service and the Sick in General Hospitals* for their anodyne, narcotic, and soporific qualities.[26] The *Confederate Receipt Book* also described an asthma cure that required inhaling the smoke from the burned leaves of jimsonweed.[27]

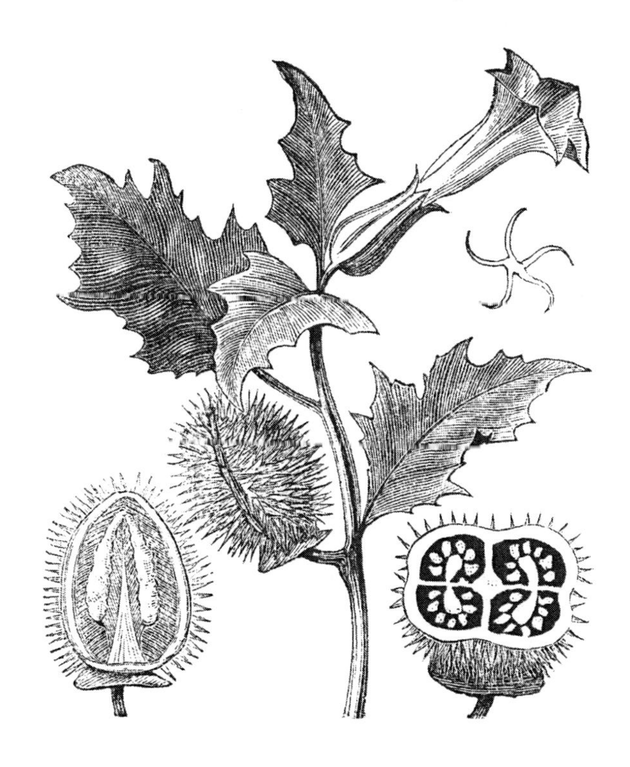

Jimsonweed was a wartime substitute for henbane imported from Europe, and both contain potent tropane alkaloids that were hallucinogenic and potentially deadly. Controlled doses of jimsonweed were used to treat pain, and the smoke was used to treat asthma (Harter, 1998).

Aside from its analgesic properties, opium was also valued for treating potentially deadly cases of dysentery. Its anti-diarrheal effects were mimicked by doses made from two common native species, wild ipecac (*Euphorbia corollata*) and hardhack (*Spiraea tomentosa*), both listed in the *Standard Supply Table of the Indigenous Remedies for Field Service and the Sick in General Hospitals.*[28] Opium remained in high demand for treating pain, so ultimately Confederate Surgeon General Samuel Moore proposed cultivating opium poppies in home gardens. His plan was to "induce the Ladies throughout the South to interest themselves in the culture of the garden poppy" and to train them in harvesting the latex for use in pharmaceutical drugs.[29] Newspapers published instructions for obtaining seed, cultivating crops, and harvesting the latex from poppy capsules. The *Yorkville Enquirer* advised incising poppy capsules at sunset because "the night dews favor the exudation of the juice, which is collected in the morning, by scraping it from the wounds. This juice is then exposed to the sun ... and kneaded from time to time until it acquires a considerable degree of consistency, when it is dried, and forms then the opium used in medicine."[30] A small quantity of opium was produced domestically, but the plants yielded low concentrations of morphine (about 4 percent) compared to the 9–14 percent concentrations from imported opium poppies.[31]

Foxglove (*Digitalis purpurea*, the source of digitoxin and other potent cardiac glycosides) was the standard treatment for dropsy, the edema associated with congestive heart failure. Unlike many other European herbs, the plants did not naturalize widely in the New World, so native substitutes were in demand. The common wildflowers known as pipsissewa (*Chimaphila umbellata* and *C. maculata*) became known as an effective and available treatment, and *C. umbellata* appeared in the *Standard Supply Table of the Indigenous Remedies for Field Service and the Sick in General Hospitals*.[32] Porcher described the species interchangeably as "aromatic, tonic, and diuretic ... easily collected around our camps, in shady woods, in almost every part of our Confederacy."[33] Other native plants with diuretic properties included twinleaf (*Jeffersonia diphylla*), tulip poplar (*Liriodendron tulipifera*), pennywort (*Hydrocotyle umbellata*), rattlesnake master (*Eryngium aquaticum*), blue flag (*Iris versicolor*), and Seneca snakeroot (*Senega polygala*).

There were also several cultivated species with purported diuretic properties, including parsley (*Petroselinum crispum*) and gourds (*Lagenaria siceraria*), as well as naturalized populations of Queen Anne's lace (the wild form of the cultivated carrot, *Daucus carota*) and water cress (*Nasturtium officinale*). Despite a long history in traditional herbalism, the European lily-of-the-valley (*Convallaria majalis*) was overlooked, although it produces potent cardiac glycosides similar to those in foxglove; the plants reproduce asexually by growth of their rhizomes and easily escaped garden cultivation. Porcher mentioned its use as an emetic, purgative, and poultice, but was apparently unaware of its use since ancient times to regulate heart activity and control edema.[34] Also overlooked were the cardiac glycosides in butterfly weed (*Asclepias tuberosa*), which was used to treat respiratory conditions and lung ailments such as pleurisy. Native Americans and early settlers had used the roots to treat dropsy, but the *Dispensatory of the United States of America* listed butterfly weed as a common treatment in southern states for pleurisy, pneumonia, and tuberculosis.[35] The *Standard Supply Table of the Indigenous Remedies for Field Service and the Sick in General Hospitals* described a decoction of butterfly weed as a diaphoretic to stimulate sweating, which practitioners of heroic medicine believed would correct humoral imbalances.[36]

Based on the Doctrine of Signatures interpretation of its branched taproot as a human torso, ginseng was regarded as a panacea for treating all human ailments.

Ginseng populations had once extended across the northern supercontinent that separated into North America and Eurasia, evolving into closely-related North American and Asian species. The *Dispensatory of the United States of America* noted that North America ginseng could be found "growing in the hilly regions of the Northern, Middle, and Western States, and preferring the shelter of thick, shady woods."[37] When Asian populations dwindled due to demand, Native Americans, explorers, and colonial settlers collected and exported American ginseng (*Panax quinquefolius*) to China. However, by the Civil War, medical confidence had waned in ginseng as a panacea, although it was still an ingredient in several proprietary tonics and drugs. Porcher encouraged investigation of ginseng properties as a stimulant, but the *Standard Supply Table of the Indigenous Remedies for Field Service and the Sick in General Hospitals* listed American ginseng only as a demulcent, capable of soothing inflamed or painful membranes.[38]

The Doctrine of Signatures interpreted the red latex in the rhizome of bloodroot (*Sanguinaria canadensis*) as useful in treating diseases and conditions related to blood and circulation, which may explain its description in nineteenth-century texts as an alterative, a drug that improves physiological function. In contrast, the *Dispensatory of the United States of America* warned of possible death from bloodroot overdoses.[39] The *Standard Supply Table of the Indigenous Remedies for Field Service and the Sick in General Hospitals* described bloodroot as a stimulant and expectorant useful in treating whooping cough, and Porcher recommended its use for lung ailments, including pneumonia and croup.[40] Folk uses included the external use of bloodroot to treat warts and ringworm fungal infections. In fact, bloodroot rhizomes contain the highly toxic alkaloid sanguinarine, which is now known for its anti-inflammatory, antimicrobial, and antiviral properties.

Skin conditions were also treated with root infusions of the hooded pitcher plant (*Sarracenia minor*), an insectivorous species with a geographic range from North Carolina to Florida. Markings on its leaves (which double as insect traps) suggested skin eruptions, and the plants became a popular folk cure in the low country of South Carolina. After the war, pitcher plants were used in the quasi-medicinal bitters made from botanical extracts preserved in alcohol, a cheap means to self-medicate but with few herbal benefits, if any. One such product was Aimar's Sarracenia or Fly-Trap Bitters, manufactured in Charleston and marketed as a cure for dyspepsia, debility, nervous complaints, sea sickness, and other ills.

Sassafras (*Sassafras albidum*) was discovered by North American explorers when the fragrant trees were mistaken for cinnamon. Both are member of the laurel family (Lauraceae) and contain terpene-based aromatic oils with similar scents and properties, including antimicrobial activity. Bark and roots were used to brew sassafras teas and elixirs that supposedly prevented aging and cured venereal diseases, and sailing ships returning to England carried sassafras as one of the earliest New World exports. Although its importance in trade diminished during the 1700s, sassafras was still widely used as a folk medicine during the Civil War. As a common southern species, sassafras was easily identified because a single tree develops aromatic leaves with four different shapes, a trait known as heterophylly. Porcher described the diverse uses of sassafras bark and oil to treat skin diseases, dysentery, syphilis, rheumatism, and other ailments, much of which also appeared in the *Dispensatory of the United States of America*.[41] Also noted was the mucilaginous nature of sassafras infusions, which seemed to prevent dehydration in the event of severe dysentery. Sassafras was also adopted in Creole cookery in which the dried leaves became known as filé powder, which was used to thicken soups and stews (see Chapter 4).

Sesame (also known as benne, *Sesamum indicum*) had similar medicinal and culinary uses (see Chapter 4). The seeds in water yielded a thin gel used to treat dysentery, credited with saving the lives of children sick with *cholera infantum* and *cholera morbus*, the chronic diarrhea that was common during the summer months and caused by poor hygiene.[42] Sesame was listed in the *Standard Supply Table of the Indigenous Remedies for Field Service and the Sick in General Hospitals*, but it was a relatively recent medicinal discovery.[43] Enslaved people had introduced sesame from Africa for its edible seeds, and Thomas Jefferson cultivated crops as a potential oilseed. The *Dispensatory of the United States of America* mentioned sesame as an emollient, laxative, and possible treatment for dysentery, noting that the plants "grow vigorously in the gardens so far north as Philadelphia."[44] Even if cultivated crops were neglected, sesame seeds were easily harvested during the war from naturalized populations. Arrowroot (*Maranta aurindinacea*) yielded another thin gel that was often used for invalid and sickroom meals. The rhizomes imported from the West Indies were difficult to obtain during the war, but experimentation with southern crops revealed that sweet potato starch dissolved in boiling water was a reasonable arrowroot substitute.[45]

Malaria, Miasma, Feverbark, and Quinine

Infections were not well understood during the Civil War, nor were the thousands of microbes that cause myriad infectious diseases. Malaria was one such vexing infection; it was often diagnosed as ague, a name derived from a French translation of *acuta*, referring to the acutely sharp fever associated with malarial infections. The unknown agents were species of the parasitic protozoan *Plasmodium*, with symptoms including intermittent fevers, chills, debility, anemia, jaundice, seizures, coma, and possible death. The vectors are female *Anopheles* mosquitoes that transmit the blood-borne disease by their bites, and once in the blood stream, *Plasmodium* protozoa enter red blood cells and reproduce. Fevers and chills occur when the infected blood cells rupture and release more *Plasmodium* into the blood. Among the species that parasitize humans, *P. falciparum* was the most severe; *P. vivax, P. ovale,* and *P. malariae* caused chronic infections with relatively milder symptoms.

None of this was known at the time of the Civil War. The nineteenth-century belief was that malaria was caused by miasma, referring to foul-smelling effluvia emitted by decomposing organic matter in stagnant water and exacerbated by warm temperatures. Agricultural journals warned of the warm weather danger of "the decay of vegetable and animal substances … generating the whole class of bilious diseases" and advised sleeping in upstairs rooms to avoid dangerous miasmatic air.[46] Yet most miasma amounted to little more than methane, the result of bacterial decomposition of plant cell walls, which are mostly cellulose. Miasma was nevertheless feared as a cause of disease, as was standing water. The connection between water and malaria was well known, as were the symptomatic chills and fevers of a malarial infection. Both were reflected in various vernacular names for malaria, which included bilious fever, intermittent fever, marsh fever, remittent fever, lake fever, chill-fever, and miasmatic fever. Malaria was also known as autumn fever because many infections occurred during the late summer and fall.

Other infectious diseases were also categorized by their presumed association with miasma. These included both "continued fevers" such as those caused by typhus and typhoid fever and "eruptive fevers" such as measles and smallpox. Typhoid became known as camp fever, a deadly bacterial infection that caused a quarter of all non-combat deaths among Confederate troops; *Salmonella typhi* was spread by body lice under conditions of poor sanitation and hygiene. Feverbark (the bark of *Cinchona* spp., the source of quinine) was adopted as a standard remedy for various fevers, even before the causative agents were understood.[47]

Miasmatic conditions were also blamed for plague, cholera, and other recurring epidemics. The miasma concept extended to open latrines, garbage piles, and associated vermin, which could indeed have spread infection. In the centuries pre-dating germ theory, disease was attributed to filth rather than microbes, and miasma was considered filthy. However, during the Civil War the French microbiologist Louis Pasteur demonstrated the connection between pathogenic bacteria and disease, as well as the impossibility of spontaneous generation of microbes from dead organic matter. By the end of the nineteenth century, germ theory and the role of insects in disease transmission were widely accepted, which led to an understanding of malaria and its spread.

Yet during the Civil War malaria was attributed to miasma and its myriad sources, from swamp mud to dirty latrines, and the possibility of miasma was considered in establishing military camps. In his role as general secretary of the U.S. Sanitary Commission, Frederick Law Olmsted published "Military Hygiene and Therapeutics" and other short publications that encouraged the careful siting of encampments to avoid miasmatic air, and if marshy habitats could not be avoided that "the ground should always ... be selected on the windward side, so that the prevailing winds should carry away the noxious emanations from the soil."[48] Regimental hospitals required 800 cubic feet of air for each patient, with effective ventilation in place and few items in patients' rooms that could absorb noxious vapors.[49]

Homes were also considered at risk. The *Report of the Commissioner of Agriculture for the Year 1863* recommended that to purify the air "from June to October a brisk fire is kindled in the family room, to burn for an hour about sunrise and sunset" and at night to avoid the low country "miasm [that] constantly rises and comes through the open windows upon the sleeper, who breathes it into his lungs corrupting and poisoning his whole blood in a night."[50] Two years later the *Report of the Commissioner of Agriculture for the Year 1865* reported that Dr. J.H. Salisbury, a New York physician and chemist, had demonstrated that malaria resulted from fungal spores that are ubiquitous in the air; he also proposed fungi as the causative agent of measles, smallpox, cholera, and plague, but his results and conclusions were deeply flawed.[51] Miasma seemed to diffuse from the swampy habitats in which the mosquitos reproduce, so it is not surprising that fungi associated with decomposition were blamed for severe fevers and chills. The *Plasmodium* protozoans that cause malaria are microscopic, and mosquito bites were so common that they were overlooked as the source of infection.

Northerners and Southerners alike feared the start of malaria season, especially in coastal areas where standing water was commonplace. Frederick Law Olmsted observed "ruinous effect upon the general constitution" of malaria, and he noted the frequency of "bilious fevers" among Northerners who visited the South during the fall months.[52] Southern planters were often infected during the rice harvest, which involved overseeing the recently drained fields where enslaved people cut and dried the sheaves.[53] Many workers (perhaps nearly half) were immune to malaria because they inherited a single gene for sickle cell anemia along with one normal gene. These heterozygous individuals did not exhibit the symptoms of sickle cell anemia, but when infected by *Plasmodium* their red blood cells quickly converted to the sickle shape and were engulfed by white blood cells. The genetic basis for resistance to malaria was not understood, but Olmsted observed that "it is dangerous for any but negroes to remain during the night in the vicinity of the swamps or rice-fields."[54] Widespread medical observations correlated climate with race and infection, noting that in the South enslaved workers were susceptible to pneumonia and other cold weather diseases, but less likely to develop severe cases of malaria.[55]

Knowledge of quinine originated centuries earlier with the use of *Cinchona* species indigenous to the Peruvian Andes. The compound is concentrated in the bark and is one of at least 24 alkaloids produced by the species. Native Americans used *Cinchona* for malaria, and during the sixteenth century they shared the cure with Jesuit priests. By the eighteenth century, feverbark imports allowed European countries to colonize the tropics without fear of malarial fevers. At the same time, explorers and colonists in North America used feverbark (also known as Jesuit bark and Peruvian bark) to prevent and treat the malarial infections that were widespread in both temperate and warm regions.

The early use of feverbark was typically as an infusion of the dried, powdered bark in wine or water. However, once French chemists isolated quinine in 1820, it was possible to prepare standardized doses. Quinine purified from feverbark became the drug of choice for treating malaria in southern states. Indeed physicians used large quinine doses to treat all fevers, including those caused by typhoid and pneumonia, a practice criticized and even ridiculed in the North.[56] Purified quinine was sold as *Quinea* or *Quiniae* and was traded primarily as an antiperiodic drug that quelled the recurring (periodic) malarial fevers. During the Civil War, shortages resulted from military demand as troops fought and bivouacked in mosquito-infested areas. Only the military could legally purchase quinine, and it was part of the standard *materia medica* carried by Union army surgeons. The federal government established processing plants in New York and Philadelphia to produce quinine extract, but Confederate states had no reliable source. Southerners thus sought quinine alternatives, which included potent extracts of garden herbs in whiskey and teas brewed from cottonseed or cayenne pepper. The latter was an old Thomsonian cure in which the spicy heat of cayenne pepper may have been interpreted by the Doctrine of Signatures as a cure for chills and fever. Some physicians believed an infusion of dried, ground cottonseed to be superior to quinine as a malaria cure.[57] A Mississippi correspondent to the *Southern Cultivator* prepared a simple cottonseed tea to treat and cure "ague and fever" by boiling a pint of seed in two pints of water.[58]

Physicians encouraged experimentation with any available substitutes for feverbark and quinine. In 1861, Joseph Jones, a Georgia doctor who became a Confederate surgeon, published a list of indigenous herbaceous and woody plants with suspected anti-malarial effects. Based on anecdotal reports and a review of medical literature, his compilation included persimmon, dogwood, black willow, catalpa, milkweed, and Virginia snakeroot. Jones encouraged colleagues to experiment with these plants and others and report on their efficacy.[59] Some practitioners instead tinkered with familiar treatments, including turpentine distilled from pine resin (see Chapter 8), which was a common counterirritant; turpentine-saturated bandages were applied to cool the intermittent fevers associated with malaria.[60] Some practitioners intuitively turned to tree bark from various species, and some sought to replicate the extremely bitter taste of quinine, contending that similar taste denoted similar medicinal use.

For those who could find it, a likely quinine substitute was Georgia bark (*Pinckneya pubens*), shrubs and small trees native to moist woodlands in Georgia, South Carolina, and Florida and related to feverbark trees in the madder family (Rubiaceae). Neither Asa Gray nor Frederick Law Olmsted seemed familiar with the species, but François André Michaux noted "its inner bark is extremely bitter, and appears to partake of the febrifuge virtues of the Cinchona for the inhabitants of the southern parts of Georgia employ it successfully in the intermittent fevers.... A handful of the bark is boiled in a quart of water till the liquid is reduced one half, and the infusion is administered to the sick."[61] Its bitter bark suggests a high alkaloid content, and Georgia bark does indeed yield compounds similar to quinine. However, even within its limited range the trees were relatively rare, so Georgia bark was not available to all who may have needed it.

The most frequently used quinine substitutes were bark harvested from dogwood (*Cornus florida*) and tulip poplar (*Liriodendron tulipifera*). Both are indigenous trees to eastern North America and grew commonly across several Confederate states, and Native Americans used both species medicinally. Neither is closely related to feverbark nor produces quinine, but Porcher extolled the value of dogwood bark as an effective substitute, noting, "This well known plant possesses tonic and anti-intermittent properties, very nearly allied to those of cinchona; in periodic fevers, one of the most valuable of our indigenous plants."[62] A commonly used bitter tonic with anti-malarial claims was compounded from the macerated bark of dogwood, tulip poplar, and black willow (*Salix nigra*) steeped in whiskey for fourteen days.[63] In fact, recent studies using dogwood extracts in alcohol reveal some activity against *Plasmodium*, perhaps explaining its widespread use and claims of efficacy in treating wartime cases of malaria.[64]

Tea brewed from boneset (also known as thoroughwort, *Eupatorium perfoliatum*) may have been somewhat effective in treating the fever and chills of both malaria and dengue fever.[65] Dengue was widely known as breakbone fever, often described as being as painful as a fractured bone. Its use was suggested by the Doctrine of Signatures; boneset has perfoliate leaves in which the blades fuse across the stem, a trait interpreted as a sign of healing. According to Porcher, "Thoroughwort or boneset tea used hot, in the cold stages of malarial fever, and cold in the hot stages, is believed by many physicians in South Carolina, who have used it since the beginning of the war, to be the very best of our indigenous antiperiodics as a substitute for quinine."[66] A correspondent to the *Confederate States Surgical and Medical Journal* also endorsed its use and suggested that practitioners collect boneset plants in September and dry the stems and leaves in a dry attic.[67] Other herbaceous plants and small shrubs used as feverbark substitutes included knotgrass (*Polygonum aviculare*), ragweed (*Ambrosia artemisiifolia*), artichoke (*Helianthus tuberosus*), hardhack (*Spiraea tomentosa*), dogbane (also known as amy root, *Apocynum cannabinum, A. pubescens*).[68] However, bark stripped from indigenous trees remained the most common feverbark alternative in the South, probably because practitioners believed that woody plants were more likely to contain quinine or similar compounds. Bark was easily recognized by species, and most importantly, supplies were available to collect in all locations during all seasons of the year.

Boneset tea was used to treat malaria and breakbone (dengue) fever and was widely regarded as an effective quinine substitute. As interpreted by the Doctrine of Signatures, the fused leaf blades are a sign of its healing properties (author's collection).

Soldiers' Medicine

The military pharmacopeia reflected soldiers' ailments, from common colds and rheumatic pains to infectious diseases, wounds, and gangrene, but routine complaints were often treated with proprietary medicines marketed directly to soldiers. Brandreth's Pills had cathartic effects and promised protection from infectious diseases; claims described them as "entirely vegetable," made from sarsaparilla (*Smilax* spp.) aloe (*Aloe vera*), gamboge (*Garcinia hanburyi*), and colocynth (*Citrullus colocynthis*). Hostetter's Stomach Bitters was an alcoholic extract of several known medicinal plants, including feverbark, centaury (*Centaurium erythraea*), sweet flag (*Acorus calamus*), anise (*Pimpinella anisum*), Virginia snakeroot (*Aristolochia serpentaria*), Culver's root (*Veronicastrum virginicum*), and the seeds of nux vomica (*Strychnos nux-vomica*), which released a small amount of strychnine into the mixture. The product was marketed as a cure-all for dyspepsia, impure blood, constipation, and malaria, but soldiers most likely purchased Hostetter's Bitters for its alcohol content, reportedly as high as 47 percent of the liquid volume.

Soldiers were also encouraged to forage for medicinal plants, in particular feverbark substitutes such as dogwood, willow, boneset, and tulip poplar (discussed above). Faced with drug shortages, the Confederate army turned to southern herbal tradition, and troops collected plants from the military pharmacopoeia for their own use, including herbs that could be brewed into tea or used as insecticides. Many had multiple uses; dried leaves from native and European pennyroyal, familiar tea herbs, prevented flea infestations in bedding. Sassafras tea was used for colds and bronchitis, and tea steeped from blackberry roots was a common treatment for diarrhea. Mayapple was useful for its laxative properties and also as a treatment for jaundice, fevers, and sexually-transmitted bacterial infections. Foraged herbs including mustards and wild onion were used to prevent and treat cases of scurvy. Farmers and planters were encouraged to save the leaves of sesame plants, which had the same mucilaginous carbohydrates as the seeds and could be used by soldiers for treating dysentery and soothing colds.[69]

Army doctors came from diverse backgrounds and favored a range of treatments for infectious diseases, wounds, and their complications. Counterirritants were commonplace; they consisted of various external treatments described in Mitchell's *Materia Medica and Therapeutics* as "the setting up of a new action in the neighborhood of a diseased spot" to divert attention away from pain and instead to skin stimulation.[70] Turpentine distilled from pine resin (see Chapter 8) was a common counterirritant, as was camphor distilled from the wood of laurel trees related to cinnamon (*Cinnamomum camphora*). Soldiers soothed their feet with counterirritant salves made from camphor, olive oil, and beeswax.[71] Another option were the poultices known as mustard plasters or sinapisms, which were prepared from ground mustard seeds (*Brassica nigra*) layered in cloth. Mustard oils, known blistering agents, caused stinging sensations and blisters

Camphor distilled from the bark of *Cinnamomum camphora* was used in counterirritant salves for military use. Camphor was also used in cotton fields to prevent the decay of cotton bolls, although its properties were more likely insecticidal rather than anti-fungal (Wikimedia Commons).

that were thought to promote healing; for the best effects, physicians were advised to grind their own seeds because commercial mustards were often adulterated.[72]

The influence of heroic medicinal practice was reflected in the imported drugs used by the military, nearly all of which had emetic or cathartic properties, including ipecacuanha, derived from the roots of *Carapichea ipecacuanha*, a coffee relative native to areas of Central and South America. The drug is an expectorant in low doses but emetic in high doses; it was combined with opium into a mixture known as Dover's powder, which was used to treat soldiers' colds. Jalap was made from the roots of *Ipomoea purga*, a sweet potato relative native to Mexico. As a strong cathartic, it was often combined with powdered rhubarb (*Rheum raphonticum*) for use as a laxative. If these drugs were not available, the most likely alternative was mayapple (*Podophyllum peltatum*, see above), also known as wild jalap and familiar to the heroic practitioners of the early nineteenth century. Squill (*Drimia maritima*) was imported from the Levant area in the western Mediterranean region and used as an expectorant and emetic. The bulbs contain cardiac glycosides that in controlled doses regulated heart activity and provided diuretic properties. Syrup of squill was prepared by boiling a vinegar extract of squill with sugar, and it was widely used in treating croup and coughs.[73] The powerful purgative known as croton oil was pressed from the seeds of the Asian shrub *Croton tiglium*, and small doses were used to treat dropsy (edema), nervous afflictions, jaundice, and cholera.[74]

Many botanical drugs are potentially lethal, and their safe use depended entirely on dose. Opium, foxglove, and squill were administered in controlled doses, as were the powdered seeds of the Asian shrub nux vomica (*Strychnos nux-vomica*), which were used to treat nervous conditions including palsy and paralysis. Minute dosing prevented poisoning from strychnine, an alkaloid that acts on the central nervous system and causes respiratory arrest. Belladonna (*Atropa belladonna*) is a perennial European herb also known as deadly nightshade. The leaves and berries contain the tropane alkaloids hyoscyamine and atropine, which are sedative in low doses and hallucinogenic and toxic at higher levels. Spread on leather, an extract of belladonna was used for relieving external pain, and it was sometimes combined with opium for its narcotic properties.[75] The European wildflower aconite (also known as wolfsbane or monkshood, *Aconitum napellus*) produces aconitine and related alkaloids that affect heart activity and can cause respiratory arrest. Carefully prepared ointments and tinctures of aconite in alcohol were used to sedate and to relieve rheumatic pain, but medical texts documented accidental deaths caused by ingesting aconite leaves.[76]

As an analgesic to counteract pain, opium was the most essential botanical drug in wartime. It is in fact a mixture of potent compounds; the latex harvested from the capsules of opium poppies (*Papaver somniferum*) contains 26 alkaloids, including morphine, codeine, laudanine, papaverine, and thebaine. The opium used to treat soldiers was imported, harvested from existing antebellum herb gardens, grown as a wartime crop, or collected from naturalized populations. Even before the war, opium was the active ingredient in proprietary soothing syrups and medicinal cordials marketed for family use, and mixtures such as Godfrey's cordial were made at home from opium, spices, and sugar. It was also a standard medical remedy for treating soldiers' fevers, bronchitis, kidney ailments, tetanus, cholera, leg ulcers, and the pain associated with gonorrhea and severe wounds. Opium was used externally in poultices applied to wounded tissue, and liniments often included it as an ingredient. Folk medicine reflects the use of opium to treat infections, and opium poultices may indeed have prevented bacteria from colonizing soldiers' wounds.

Dosing involved variable concentrations of potent opium alkaloids, long known to be potentially dangerous; centuries earlier Hippocrates and Galen advised care in administering opium because it could cause death during a sleep-like state. However, in 1804 the German chemist Friedrich Wilhelm Sertürner isolated morphine from opium (the first such purified botanical alkaloid), which he named for Morpheus, the Greek god of dreams. As a powerful analgesic, pure morphine allowed for more accurate dosing, although opium was still stocked by Civil War pharmacies and used in tinctures, Dover's powder, and other compounded drugs. Both opium and morphine were used to treat pain from battle wounds, both acute and chronic, and huge amounts were consumed. The Union army provided nearly ten million opiate pills to troops, in addition to nearly almost million ounces of various opiate-based tinctures and powders.

Civil War surgeons also valued opium as a hemostat essential in staunching blood loss from wounds. As noted in *A Manual of Military Surgery* (1863), "Opium is the one indispensable drug on the battlefield—important to the surgeon, as gunpowder to the ordnance officer—for besides the ages of pain for which it is the reprieve as an anodyne, it saves rivers of blood as a haemostatic."[77] It was often used alongside ergot (*Claviceps purpurea,* known medicinally during the nineteenth century as *Secale cornutum*), a parasitic fungus that infects rye and wheat. Controlled doses of ergot were used to stop hemorrhages by constricting peripheral vessels, but toxic doses resulted in gangrene from loss of blood flow. The fungus produces an alkaloid that includes lysergic acid as part of its chemical structure; flour made from tainted grain also caused muscle spasms and hallucinations.

The introduction of hypodermic syringes during the late 1850s enabled precise dosing, and physicians and patients began to regard morphine as a panacea. In addition to injections, larger syringes were used to administer analgesic enemas containing combinations of morphine, belladonna, and henbane; these devices were molded from gutta percha, a thermoplastic latex harvested from *Palaquium percha,* trees native to Malaya.[78]

Antebellum medical texts warned of the habit of opium-eating and included graphic descriptions of addiction: "It is a vice that hurls from the loftiest elevation the most towering intellect; dissipates into thin vapor the solid lore of ages; transforms the giant into the veriest pigmy; sinks lower than the brute the being made in the image of his God. And thus works this horrible vice."[79] Yet by the end of the war, accidental cases of opium overdoses and addiction were alarmingly common among both men and women, including enfeebled veterans and exhausted civilians. Overuse of opium and purified morphine spawned addiction, which became known as a soldiers' disease associated with lifelong dependency. Tannin and strong coffee were considered possible opium antidotes, although overdoses were typically treated with emetics including ipecacuanha and tartar emetic.[80] Morphine syringes became commonplace, and proprietary opiate mixtures proliferated until the Harrison Narcotic

Act of 1914 established legal controls of addictive drugs.[81] Addiction was overcome with nothing more than willpower, and the misery of opiate dependency encouraged experimentation with supposedly non-addictive alternatives. John Stith Pemberton, a Confederate lieutenant colonel who became addicted to morphine, combined cocaine (an alkaloid derived from the leaves of coca plants [*Erythroxylum coca*] and caffeine extracts of kola nuts [*Kola* spp.]). His "brain tonic" evolved into the modern beverage known as Coca-Cola.

Whiskey was used medicinally in army hospitals, but desperate civilians believed that grain was needed for bread rather than fermentation and distillation. The Confederacy produced medicinal whiskey at a large distillery in Salisbury, North Carolina, despite severe food shortages and the tendency for distilleries to increase local grain prices. Some farmers refused to pay their in-kind tithe tax (see Chapter 3) unless the Confederate government promised that the grain would not be used in making whiskey.[82] Barrels of alcohol imported from Europe were a wartime substitute that conserved corn and other grains for use as food.[83]

Alcoholic stimulants were prescribed for typhoid symptoms and the deep abscesses described in medical texts as phlegmonous erysipelas, now known to be an infection caused by *Streptococcus pyogenes*. Medicinal beverages included milk punch made from whiskey and whey made by coagulating milk with wine; brandy, a liquor made by distilling wine, was widely prescribed for soldiers suffering from shock and collapse.[84] As noted in *A Manual of Military Surgery*, "Among the many stimulants with which the surgeon's armamentarium is supplied, there are few more acceptable to the stomach than brandy, and none as well calculated to fulfil the indications which here present themselves."[85] A combination of brandy and opium was used to relieve the muscle spasms associated with tetanus, infections caused by common soil bacteria (*Clostridium tetani*) that colonized deep wounds.[86]

Ingenuity remained essential, especially in the South where wartime shortages were severe. Life preservers made from the bark of cork oak (*Quercus suber*) were repurposed to seal bottles of bitters and medicines, and substitutes were made from corncobs and rags. Tourniquets were made from strips of bark, fence rails were fashioned into splints, and injured limbs were positioned on pillows filled with straw or oak chaff. Bandages were generally in short supply, and used cotton and linen bedding was repurposed to use in army hospitals; the natural oils on unprocessed cotton fibers prevented absorption of the blood and pus associated with wounds.[87] Soap was made from potash (typically collected from the ashes of oak and hickory wood) and any available fats, including oils pressed from cottonseed, castor beans, and the seeds of chinaberry trees (*Melia azedarach*).[88] The latter were planted as ornamental trees along the avenues of many plantations (see Chapter 2), but their widespread use for wartime soap-making was unanticipated.

Slavery and Medicine

The botanical medicine practiced by enslaved people evolved from diverse sources, including colonial herb uses, southern folk wisdom, Native American healers, and African and Caribbean traditions. Knowledge of medicinal herbs was part of the oral tradition in families, and enslaved women often had expertise with healing that exceeded the medical knowledge of slaveholders and their families. However, medicinal plants are often toxic in uncontrolled doses, and overseers were sometimes suspicious of enslaved people with medical expertise, lest they use their knowledge of practical chemistry to prepare poisons. Thus colonial laws in Virginia, South Carolina, and Georgia forbid enslaved people from preparing medicines, at least those that might be toxic.[89] At the same time, enslaved people were sometimes suspicious of trained physicians and conventional medicine, perhaps stemming from the antebellum hospitals and infirmaries for enslaved people that served the interests of slaveholders who paid for their care. Such facilities varied in their standards of practice and hygiene, and they were often distrusted by patients.[90]

Enslaved people had monetary value, but conditions like mental illness and epilepsy could interfere with their sale at auction. These conditions were often commingled and confused, and standard drugs included tartar emetic and rhubarb, both familiar to nineteenth-century practitioners of heroic medicine.[91] Planters were also concerned with infectious diseases that could affect enslaved people who lived in close quarters. A treatise on cholera titled *Observations on the Epidemic Now Prevailing in the City of New-York* (1832) advised southern slaveholders to use jalap (a purgative drug from the roots of *Ipomoea purga*), laudanum (a tincture of opium), and tartar emetic (a salt of tartaric acid from fermented grapes) to treat their workers, all standard treatments at the time.[92] Manufacturers of proprietary medicines saw an opportunity to market directly to slaveholders concerned with disease caused by crowding and poor nutrition. Swaim's Panacea promised to cure dysentery, venereal diseases, parasitic infections, and the scrofulous sores caused by tuberculosis which spread in close quarters. Swaim asserted that "Negroes who are confined in large numbers on plantations in hot climates, are particularly liable to such forms of disease, arising from a vitiated state of the blood and want of cleanliness and variety of food."[93] The two primary ingredients of the mixture were oil of wintergreen (methyl salicylate isolated from wintergreen, *Gaultheria procumbens*) and sarsaparilla (*Smilax ornata*), a North American species that became known in Europe as a cure for syphilis.

For the most part, however, enslaved people were responsible for their own care and cures, and they followed an empirical approach that depended on experience, observation, and acquired practical knowledge. European herbs such as tansy, horehound, catnip, pennyroyal, and sage were brewed as teas to treat colds and other common ailments, and enslaved people eventually cultivated many of these plants as accessible remedies.[94] All are now known to contain terpene-based essential oils with antibiotic properties. Tea was also brewed from comfrey (*Symphytum officinale*), but it is now known to contain potentially toxic pyrrolizidine alkaloids as well as allantoin, which promotes wound healing.[95] Cherokee people shared knowledge of butterfly weed, a source of potent cardiac glycosides that they used for treating pleurisy, a use later mentioned in several of the slave narratives recorded by the Works Progress Administration (WPA) during the 1930s.[96] The Cherokee practice of using pokeweed in salves and poultices to treat ulcerous sores and warts was mirrored in the use of the roots by enslaved people for preparing liniments for bruises and sores.[97] However, recent analysis reveals that pokeweed roots contain potentially toxic mitagens (proteins that stimulate cell division), and handling the plants should be avoided. Other North American plants adopted by enslaved people included dogwood and boneset for treating fevers, knowledge probably acquired from slaveholders who sought alternatives to imported feverbark.

In particular, the WPA narratives include numerous references to asafetida (*Ferula assa-foetida*), a perennial herb related to Queen Anne's lace (also known as wild carrot, *Daucus carota*, family Apiaceae), which originated in the deserts of Iran and Asia. Asafetida was recognized in ancient times as a medicinal plant and was introduced to Africa, where local people adopted the pungent roots for treating diverse maladies. Enslaved people probably carried asafetida to America, where its diverse uses in both medicine and magic continued. The roots contain a complex oleoresin gum that includes disulfides and polysulphanes; this chemistry explains the frequent association of asafetida with the scent of onions, which produce similar sulfurous compounds.

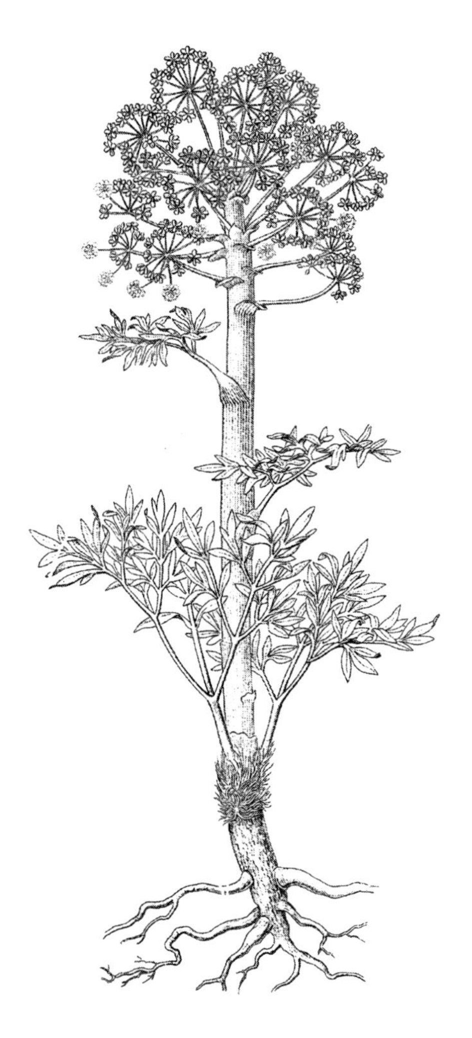

Enslaved people introduced asafetida to America, where the plants were cultivated as medicinal herbs and used to ward off contagious childhood diseases. The plants are related to silphium which Greek and Roman women used as an abortifacient; enslaved women used decoctions of asafetida to control birth rates (Harter, 1998).

The foul scent and taste of asafetida enabled easy identification of the plants and probably suggested a tradition that resembled the wearing of garlic to protect against disease. Magic and medicine were practiced together, and practitioners and patients placed faith in both botanical cures and charms. The spiritual tradition known as rootwork pivoted on the use of asafetida for its presumed magical properties, encompassing numerous practices that were widespread among enslaved people in southern states during the mid-nineteenth century. Enslaved workers placed faith in red flannel bags containing asafetida roots, red pepper, sassafras, and snakeroot to provide charms and spells, cure afflictions, and solve problems. Enslaved children wore similar bags of asafetida to ward off measles, diphtheria, pertussis, and other contagious diseases. The bags were tucked under clothing out of sight of disapproving masters, and the blood red color of the cloth was thought to enhance the effectiveness of the plant material in rendering magic.[98]

Asafoetida roots were used internally for their laxative and emetic properties and to promote digestion. In Africa, asafetida was a likely substitute for silphium, a closely related North African species used by local people for various pains, coughs, and fevers. Ancient Greek and Roman women used silphium as an abortifacient, and demand for the plant apparently caused its premature extinction before it was recorded and named botanically. Asafetida had a similar use in folk medicine as an abortifacient, and decoctions of the roots may have caused miscarriages in enslaved women who did not ingest the herb with the intention to abort a pregnancy. The active principles were probably coumarins similar or identical to ferujol, which has been isolated from an Asian species of *Ferula* used as an abortifacient in early pregnancies.[99]

Some historians have interpreted contraception and abortion as a means by which enslaved women resisted the control of masters and overseers, perhaps by avoiding the birth of children who might be sold. Four or five years frequently passed between births, which probably cannot be explained by long periods of lactation.[100] In addition to asafetida, the contraceptive chemistry of cotton (*Gossypium* spp.) was realized in Africa, where women chewed the roots of tree cotton (*G. arboreum*) to prevent births by aborting early pregnancies. The active agent is gossypol, a polyphenolic cotton pigment which in nature discourages herbivorous insects. Gossypol is now known to cause developmental disorders; ingestion may have affected early fetal development, which was the likely basis of its use as an abortifacient. Plantations were often a ready source of cotton roots that women could secretively collect and store. The WPA narratives describe the foraging of the roots of cultivated cotton (*G. hirsutum* and *G. barbadense*, see Chapter 3) by women who wanted to prevent or abort pregnancies.[101] The irony cannot be overlooked that the crop that created a need for enslaved labor also provided a means for women to control pregnancies and suppress familial slavery. Low concentrations of gossypol also stop spermatogenesis, and it is possible that males may also have ingested sufficient gossypol to prevent conception; in fact, some recent studies have centered on gossypol as a possible male contraceptive drug.[102]

Other plants used by enslaved people for contraception included indigo, including both the cultivated species *Indigofera tinctoria* and possibly the native species *I. caroliniana* and *I. leptosepala*. The abortifacient effects of indigo likely result from the presence of indospicene, a unique amino acid known to cause damage to developing fetuses. Root decoctions of Bengal indigo (*I. arrecta*), native to Africa and also used as a blue dye, were used as abortifacients in traditional medicine, and it seems likely that enslaved people found indigo in America and adopted it for the same use. Other botanical drugs to control births included emmenagogues prepared from the roots of fennel (*Foeniculum vulgare*, referred to as dog fennel in the WPA narratives) and the widely dispersed aquatic weed known as red shank (*Persicaria hydropiper*, also known as *Polygonum hydropiper*). Sometimes these were combined with a dose of turpentine and blackhaw (*Viburnum prunifolium*). Blackhaw was used by Native Americans for treating female problems, including the prevention of miscarriages, contradictory to its use by enslaved women as a possible abortifacient. In contrast, slaveholders often dosed women with infusions of blackhaw to encourage healthy pregnancies, perhaps following Native American tradition. Dose may have been the determining factor. The complex chemistry of blackhaw includes salicin, an aspirin precursor that may have been useful in relieving cramps, spasms, and pain; ingesting blackhaw also seems to relax uterine muscle, perhaps the result of the glycoside scoploletin.

6

Gardens and Horticulture

Plantation Gardens

Plantation gardens reflected the needs of households and the desire for pleasure, and horticultural specialization was possible with so much available land, typically 500–1,000 acres. Kitchen gardens yielded vegetables, fruit, and herbs for rural homesteads, while pleasure gardens were landscaped for the prestige that they offered. However, horticultural reality was often a struggle; plantation owners desired Old World design, but standard farm and garden texts were largely adapted from European sources and provided little guidance for planting and growing in the Deep South.

Plantations typically had a long growing season, high humidity, warm climate, and seasonal drought, geographic conditions first addressed by Francis S. Holmes in *The Southern Farmer and Market Gardener* (1842).[1] Gardens had to accommodate plants that barely tolerated these conditions, as well as semi-tropical species that thrived in southern heat and humidity. Organized as a series of topical articles, Holmes' test provided useful information for southern kitchen gardens, from growing figs to propagating sweet potatoes and planting successive pea crops. William Nathaniel White discussed domestic landscapes in *Gardening for the South* (1856), but this was published years after the pleasure gardens of most antebellum plantations were designed and installed.[2] He suggested decorative plantings within view of the home, but his vision was comparatively modest compared to the grand gardens on many large plantations.

Plantation gardens and houses reflected a preference for symmetry and balance, a style that took root in the early Southern colonies and continued through the antebellum years on an increasingly larger scale, as land and wealth permitted. Pleasure gardens conveyed social status because their installation and ongoing maintenance were affordable only to the rich. Specimen trees and shrubs could be admired from all sides rather than partially camouflaged in a semi-natural landscape. On a plantation scale, pleasure gardens used combinations of formal geometric parterres and pathways, colorful flower beds, and successive flowering times. Geometric parterres and pathways denoted control of nature but demanded intensive labor; heat and humidity stimulated aggressive growth, so parterres required frequent tidying, clipping, and weeding. Southern planters resisted change even as landscape gardens became popular in England, although some preserved groves of native trees on their land probably because they provided shade and timber. Southerners ignored Andrew Jackson Downing's persuasive arguments for planned and cultivated woody landscapes described in his 1841 *Treatise on the Theory and Practice of Landscape Gardening* (see Chapter 2). According to the *Southern Cultivator*, "On most farms, land enough is lying waste, to make a picturesque landscape, at a small expense" by planting trees in natural groupings, planning curved pathways, and installing rustic arbors and bridges.[3] However, southern tastes leaned toward formality and symmetry.

Parterre beds were edged with dwarf boxwood hedges (*Buxus sempervirens* var. *suffruticosa*) and planted seasonally with flowering bulbs and herbaceous plants with desirable color and scent. Boxwood had favorable traits, including evergreen foliage, sun and shade tolerance, and dense branching. Despite their slow growth, dwarf boxwood hedges required trimming to maintain their height at eight to ten inches; gardeners were also advised to prune the roots, a technique resembling bonsai that promoted compact growth. Planting hundreds of seedlings to fill parterre beds was an undertaking that required a glasshouse and labor. Early bedding plants included pansy (*Viola* X *wittrockiana*), larkspur (*Delphinium* spp.), dahlia (*Dahlia* spp.), hollyhock (*Althaea rosea*), stock (*Matthiola incana*), and gillyflower (*Dianthus caryophyllus*), herbaceous annuals and perennials that would have been at home in an English cottage garden.

Dwarf boxwood was used to edge the elaborate parterre gardens; the labor-intensive work included pruning both the shoots and roots to promote compact growth. Boxwood hedges surrounded colorful pansies, dahlias, hollyhock, or other herbaceous and perennial bedding plants (morphart © 123rf.com).

Tall species required labor-intensive staking, but plant selection ultimately depended on survival rather than attributes or origins. Thus by the 1860s, parterres included tropical annuals such as portulaca (*Portulaca oleracea*) and petunia (*Petunia* hybrids), which provided continuous bright color during a long, hot growing season, in comparison to many European flowering plants that withered in extreme heat.

Planters and their families took an interest in finding new options among recent botanical introductions, both herbaceous and woody. As a matter of routine, journals such as the *Southern Cultivator* mailed the seeds of recent introductions to their readers, and they printed instructions for sowing and cultivation in their monthly publications.[4] Ornamental Asian shrubs provided impressive color, scent, and evergreen foliage in southern habitats, but some introductions naturalized quickly and in some cases became invasive (see Chapter 2). Some Northerners grew camellias (*Camellia* spp.), magnolias (*Magnolia* spp.), and Japanese pittosporum (*Pittosporum tobira*) as hothouse specimens, but Southern planters soon realized through trial and error that many woody plants from Asia survived year round outdoors in parterre gardens.

Some members of the planter class travelled abroad and returned home with plans of hiring gardeners capable of planning and planting formal European-style gardens. Others simply desired fashionable plantings, but soon realized that parterre gardens required expertise in both design and horticulture. Some parterres were the work of Americans who relied on texts such as *The Retired Gard'ner* (1706), an English translation of two earlier French books on formal gardens and their design.[5] The combined volumes included diagrams and instructions for selecting a garden site, levelling and improving soil, spreading gravel on pathways, and using trees, shrubs, grass plots, flowers, and potted plants to achieve particular symmetry and effects. Designs included squares, rectangles, triangles, diamonds, circles, and even teardrop shapes, and parterre styles were named for textile designs and included "cutwork" and "imbroidery." Once outlined in the soil, shapes were defined using dwarf boxwood to edge the beds. This important step demanded expertise (or at least patience) and established the parterre for years to come. *The Retired Gard'ner* described the close planting of boxwood cuttings to create a desirable unbroken network of low hedges.

Many parterre gardens were the work of skilled itinerant horticultural designers from England, Scotland, Ireland, and France who travelled to southern states to ply their craft. They quickly adapted their knowledge and practice to southern growing conditions. Advertisements appeared in journals and newspapers offering services by the month for garden planning, planting, improvement, enlargement, and repair. Word of mouth and successful results increased demand for the services of itinerant practitioners. In short, the gardens were meant to be seen and admired, which worked to the advantage of those who could establish and maintain them. Many plantations were located on rivers, so pleasure gardens featuring parterre designs were intentionally sited with clear views from the water, with land access provided by a tree-lined drive.

Although southern pleasure gardens were bankrolled by nineteenth-century cash crops, they reflected eighteenth-century design and conservative southern values. The demise of European-style plantation horticulture came with wartime struggles and the fall of the Confederacy. As gardeners were called to military service and enslaved workers were freed, there was insufficient labor to maintain extensive grounds and gardens. No doubt southern gardens were naturally prone to excessive heat and humidity, drought, insect pests, untrammeled growth, falling debris, and weeds. Cultivated areas were quickly overtaken by invasive non-native plants, most of them Asian imports introduced for garden cultivation. Indeed it seems unlikely that southern pleasure gardens ever achieved the picture-perfect ideal of formal gardens cultivated in temperate European climates. Even in the years leading up to the war, travelers chronicled the collapse of plantation economy and deterioration of the cultivated landscapes and gardens. In *A Journey in the Back Country* (1860), Frederick Law Olmsted documented abandoned plantations, often the result of depleted soil, which foreshadowed the demise of the plantation system of agriculture and horticulture.[6] John Greenleaf Whittier, a poet and Abolitionist, described southern decay and deterioration in his anti-slavery poem "The Panorama" (1856):

> A slave plantation's slovenly repose....
> In shabby keeping with his half-tilled lands;
> The gates unhinged, the yard with weeds unclean.

Union troops raided southern plantations during the latter years of the war (see Chapter 2), destroying fences and pilfering anything of value. Accounts of raids describe the theft of roses and evergreen shrubs, the burning of trees, and the deliberate destruction of gardens.[7] Freedmen became tenant farmers, farm workers, and sharecroppers, and they no longer provided the free labor that maintained cultivated plantation landscapes. In northern states, domestic horticulture also suffered, albeit to a lesser extent, as noted in *The Magazine of Horticulture*: "Cultivators have been called from the garden to the battlefield, and the spade has been changed for the sword.... May we hope that those still in the field may return to the pleasant gardens and happy homes that they have so promptly left at their country's call."[8]

Botanical Origins

Since their colonial origins, formal plantings typified antebellum gardens, a drastic contrast with the asymmetrically natural plantings favored in England and advocated for American gardens by Andrew Jackson Downing (see Chapter 2). The earliest southern parterres were edged with native plants, which were subsequently introduced to English gardens as worthy of cultivation. Although they were planted as ornamental trees and shrubs, several had medicinal properties that may have been useful in wartime homes. American holly (*Ilex opaca*) and yaupon holly (*I. vomitoria*) had evergreen foliage and long lasting red berries on the pistillate (female) shrubs. Yaupon leaves lack the spiny points found on the foliage of American and English holly, and the shrubs were planted as low parterre hedges in areas where boxwood succumbed to high humidity.

Like English holly (*I. aquifolium*), both American hollies could also be grown as tall hedges, specimen plants, or topiaries, and they had similar medicinal uses in folk medicine for treating colic and dropsy. Native Americans used yaupon holly to brew emetic, high caffeine infusions used ceremonially for purification. Daniel Solander, a student of Linnaeus, apparently knew of this property when he applied the epithet *vomitoria*, although milder infusions became a colonial substitute for Asian tea and an American export to England during the eighteenth century. However, American holly was notoriously slow growing and difficult to transplant; in the *Ladies' Southern Florist* (1860), Mary Rion suggested digging small hollies from the forest "just before the buds begin to shoot. The smaller the plant, the better the success in transplanting. Protect them a long time from the sun's rays."[9]

Cherry laurel (*Prunus caroliniana*, also known as wild orange) was another valued native evergreen, a cherry with foliage reminiscent of bay laurel (*Laurus nobilis*). Its crushed leaves release a scent resembling almond extract, produced when cardiac glycosides release hydrogen cyanide in damaged tissue. Whether planters realized its potential toxicity is unknown, but deer avoid browsing on the leaves. Mary Rion described the species as "one of the most beautiful vegetable productions of the South" and suggested it for tall hedges or single specimen trees.[10] Cherry laurel seeds were slow to germinate, but the seedlings grew quickly and were valued by planters who wanted both parterre gardens and small trees to shade their homesteads.

Sweet shrub (*Calycanthus floridus*, also known as Carolina allspice) was also known for its characteristic scent, in this case arising from an essential oil aromatically reminiscent of cinnamon and allspice. By the mid–1800s, sweet shrub was widely advertised in nursery catalogues in both the North and the South, described as providing a pleasant scent to an entire garden. The shrubs produced deep reddish-brown flowers, rarely grew more than about five feet in height, and thrived in garden habitats. Fringe tree (*Chionanthus virginicus*) merited attention for its white flowers in the spring, followed by deep blue olive-like fruit in the fall. It was valued in antebellum gardens as one of the earliest spring flowering woody plants. The trees were sometimes transplanted from wild sites or started from seed, but germination took months to occur. Mary Rion observed that fringe tree "does best when grafted on the common ash," and early gardeners experimented with grafting fringe tree branches onto ash saplings (*Fraxinus americana*; both species are in the olive family, Oleaceae).[11] Native

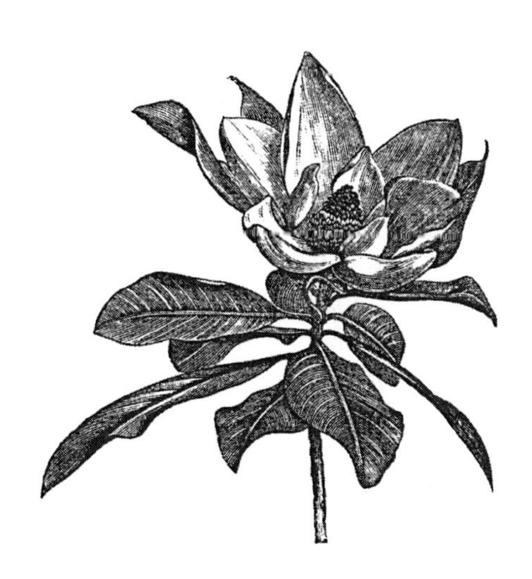

Southern magnolias were valued for their evergreen leaves and scented flowers, regarded by many as the finest woody plant discovered in North America. They were planted widely on plantations and used for medicinal tinctures and decoctions during the Civil War (author's collection).

Americans poulticed the bark to treat infections, and some settlers used it as a folk substitute for feverbark (see Chapter 5) in preventing or curing so-called intermittent fevers or malaria.

The most valued native ornamental plant was the southern magnolia (*Magnolia grandiflora*), which combined the attributes of evergreen foliage and flowers with saucer-like size and a fruity scent, traits that evolved as adaptations that attract beetle pollinators. The size of southern magnolias depended on soil quality, and some reached heights of 100 feet in ideal plantation habitats. Many botanists and nurserymen considered the southern magnolia to be the finest woody plant in North America, and it competed successfully with the Asian imports that flooded American nurseries during the 1840s. For many, the southern magnolia was the botanical symbol of the South, as described in 1861 by Confederate General Albert Pike of Arkansas in the lyrics to his song "The Magnolia":

> What, what is the true Southern Symbol,
> The Symbol of Honor and Right,
> The Emblem that suits a brave people
> In arms against number and might!—
>
> 'Tis the ever green stately Magnolia,
> Its pearl-flowers pure as the Truth,
> Defiant of tempest and lightning,
> Its life a perpetual youth.
>
> French blood stained with glory the Lilies,
> While centuries marched to their grave;
> And over bold Scot and gay Irish
> The Thistle and Shamrock yet wave:
>
> Ours, ours be the noble Magnolia,
> That only on Southern soil grows
> The Symbol of life everlasting;—
> Dear to us as to England the Rose.

Southern magnolia trees also had practical medicinal uses. Native Americans used a decoction of the aromatic bark to treat dropsy and skin conditions and leaf decoctions for colds and chills. Other uses included tinctures prepared from the bark or woody fruits (follicles) to treat fevers. In *Resources of the Southern Fields and Forests* (1863), Frances Peyre Porcher suggested various uses for southern magnolias in treating rheumatic pains, gout, "pectoral affections," and other ailments, and faced with drug shortages, planters' families may have harvested leaves from garden trees for household use.[12] Sesquiterpene lactones and other phytochemicals in southern magnolias are now known to have antimicrobial properties, which may explain their efficacy in treating some infections.

Tulip poplar (*Liriodendron tulipifera*, a magnolia relative in the family Magnoliaceae) was known for its fast growth and showy yellow-green flowers with orange markings. George Kidd, an English landscape gardener based in Columbus, Georgia, suggested planting tulip poplars on lawns; he favored native species arranged in natural plantings, noting in correspondence to the *American Cotton Planter and Soil of the South*, "I agree with you, that in the improving of the grounds exotics can very well be dispensed with, seeing that we have such a vast variety of highly ornamental trees and shrubs indigenous to our soil."[13] However, he realized that exotic nursery plants and recent introductions held interest for many and that few Americans would invest highly in gardens limited to native species. Formal southern gardens reflected horticultural demand for showy and varied cultivated specimens, and the notion of exotic plants became so commonplace that it was used to advertise a collection of popular sheet music titled "The Exotics—Flowers of Song Transplanted to Southern Soil," published during the 1860s by A.E. Blackmar in Augusta, Georgia. The title page was illuminated with various twining vines, foliage, and flowers.

By the mid-nineteenth century, nurseries propagated and marketed many ornamental plants discovered by botanists and explorers in China and Japan, and these plants thrived in American gardens. In terms of plant geography, this was not surprising; there are about 120 genera with closely related species in eastern North America and eastern Asia. They are remnants of the vast forests that once grew across the northern continent of Laurasia before the Atlantic Ocean separated the landmass into North America and Eurasia. Garden genera such as magnolia *(Magnolia)*, dogwood (*Cornus*), hydrangea (*Hydrangea*), and maple (*Acer*) have natural distributions in eastern Asia and eastern North America, and it was predictable that Asian imports, many of which are comparatively showier and faster growing, would become popular in the American nursery trade.

With the land and money to underwrite large pleasure gardens, by the 1840s Southern planters were buying large numbers of Asian cultivars. Magnolias included the lily-flowered magnolia (*Magnolia liliiflora*) and the Yulan magnolia (*M. denudata*), both native to China. The saucer magnolia (*M.* X *soulangeana*) is a hybrid of these two species. It was first listed in the *1831 Annual Catalogue of Fruit and Ornamental Trees and Plants* published by the William R. Prince botanic garden and nursery in Flushing, New York, and became known for its pink or purplish flowers that appear each spring before the foliage. Northern gardeners were advised that Asian magnolias "succeed very well in sheltered situations, in our pleasure-grounds, and add greatly to their beauty early in the season," while southern gardeners planted them in parterres as specimen trees with ornamental flowers and foliage.[14] In contrast, the attraction of the evergreen Chinese magnolia known as banana shrub (*M. figo*, earlier classified as *Michelia figo*) was primarily its fruity scent. The flowers release isoamyl alcohol, the precursor of banana oil (isoamyl acetate), and women carried them as a source of perfume. According to the *Southern Cultivator*, the flowers were "pale straw color, each petal edged with carmine, about 2 inches in diameter and exquisitely fragrant, like a ripe banana, whence its name. As yet quite rare."[15]

Camellias (*Camellia japonica*) became known as an iconic southern plant although the species originated in China, Japan, and Korea. Their waxy evergreen leaves and floral variations made the plants quite appealing for viewing in parterre beds; cultivars included single and double flowering varieties and a range of red, pink, and white petal colors. Asian camellias first appeared in the *1807 Catalogue of Trees, Shrubs, and Herbaceous Plants* offered by William Bartram in Philadelphia, and they were described initially as hothouse plants. However, southern gardeners soon discovered that camellias survived mild winters, and by mid-century the shrubs and small trees grew in plantation gardens as hedges and specimen plants. Tea (*C. sinensis*) is a closely related Chinese and Indian species which some believed would also thrive in southern climates, but attempts to establish a successful American tea industry in Confederate states failed as a result of wartime strife and competition from tea imported from Asia (see Chapter 4).

Other Asian species with showy flowers included the small trees of crepe myrtle (*Lagerstroemia indica*), a heat tolerant Chinese species with white, pink, purple, and red cultivars. Although they did not flower reliably in cool climates, crepe myrtle imported from English nurseries produced dramatic floral displays in southern states. The trees were easily propagated from shoots that grew from the shallow roots, and during the early nineteenth-century crepe myrtle specimens were planted in many parterre beds, where they thrived in both acidic and alkaline soils. Showy flowers were also the attraction of royal paulownia (*Paulownia tomentosa*) from China, first introduced to European gardens in the 1830s. The catalogue of the William R. Prince botanic garden and nursery in Flushing, New York, noted that the trees bore "a profusion of beautiful purplish bell-shaped flowers" resembling foxglove, with foliage similar to catalpa.[16] Gardeners planted the fast-growing trees in pleasure gardens and parterres where they produced large numbers of wind-dispersed seeds. Royal paulownia tree seedlings do not tolerate deep shade, but they germinated and thrived widely in abandoned fields and disturbed sites, soon becoming invasive (see Chapter 2). In the years following the Civil War, royal paulownia trees were regarded as horticulturally coarse and rarely cultivated.

Royal paulownia trees were introduced from China for their showy purple flowers, and fast-growing saplings were widely planted in plantation gardens. The wind-dispersed seeds germinated in waste sites, and the trees were soon regarded as undesirable weeds (author's collection).

The shrubs of Japanese pittosporum (*Pittosporum tobira*) from China and Japan have evergreen foliage (including a variegated form) and clusters of sweet-scented white flowers in spring. The leaves tolerate salt spray and thus became popular in formal gardens in coastal areas of the Deep South, including Savannah and New Orleans. Mimosa (*Albizia julibrissin*) was somewhat of a curiosity, with elongated pink stamens that replaced the attractant function of pigmented petals. The Asian trees bear bi-pinnate leaves (pinnately compound leaves in which the leaflets are also pinnately compound) and grew quickly into mature specimen trees in formal gardens. As with other legumes, the roots of mimosa trees have a symbiotic relationship with nitrogen-fixing bacteria (see Chapter 2), and its seedlings easily naturalized in areas with nutrient-poor soils.

In some cases, remarkable foliage was the main attraction for garden cultivation. Chinese parasol trees (*Firmiana simplex*) have palmate leaves ten to twelve inches in diameter, and the mature trees develop an umbrella-like crown. The shrubs of Japanese aucuba (*Aucuba japonica*) have evergreen foliage variegated with bright yellow markings, and they tolerate summer heat, described as "very remarkable for its maculated road foliage; stands the sun quite well."[17] Tree of heaven (*Ailanthus altissima*) is a Chinese import with large compound leaves, and it was prized for it fast growth into a mature specimen. Seedlings grow three to six feet annually and produce offsets which are clones of the parent tree, but many gardeners soon tired of its noxious scent and invasive growth in fields. During the years of Reconstruction, farmers observed that cows that ate tree of heaven leaves seemed immune to undiagnosed bacterial infections including anthrax and erysipelas, diseases known generally as murrain. Tree of heaven was described as a "sure preventative of murrain in cattle. The cattle commence eating the leaves of the tree about the time in the season when murrain would appear; and none having access to it have ever been known to take the disease, while others all around have been seriously afflicted."[18] Physicians prescribed resin from tree of heaven for its emetic properties, and leaf and bark infusions were used to treat dysentery, malaria, and parasitic infections. Its antimicrobial properties seem to result from the quassinoid compounds that occur in tree of heaven and related species (family Simaroubaceae). In the soil, its phytochemicals slow the root growth of competing plants (a survival adaptation known as allelopathy), which contributes to its invasive growth in abandoned fields.

Mid-nineteenth-century parterres were created with a geographical mix of European garden plants, species native to North America, and recent Asian introductions. Colonists carried traditional garden plants from England, including mock orange (*Philadelphus coronarius*), oleander (*Nerium oleander*), chaste tree (*Vitex agnus-castus*), and boxwood (*Buxus sempervirens*), which formed the woody framework of southern colonial gardens. By the 1850s, some planters abandoned boxwood in favor of an Asian evergreen shrub, replacing traditional hedges with *Euonymus japonicus*, now known commonly as the Japanese spindle tree. Its leaves are shiny and deep green, about twice the size of boxwood leaves, and according to Mary Rion, "it grows well from cuttings. No necessity for small plants to have roots, as they will grow without."[19] In other words, plants rooted easily from shoots merely thrust into the soil, a trait that must have appealed to gardeners. The shrubs became known as ideal for southern hedges, but spindle tree seeds are each surrounded by a red-orange aril that attracts hungry birds and results in wide dispersal. Predictably the shrubs soon naturalized, and some populations became locally invasive.

The same attributes that made many Asian plants easy to cultivate and propagate also resulted in tendencies toward invasive growth. Many Asian species grew fast in warm, moist climates, often exhibiting vast root growth and the tendency to clone and colonize land by producing offsets (suckers) of the parent plant. Seed production was typically high, sometimes enhanced by bird dispersal and high germination rates. Indeed some of the most aggressive invasive plants in North America originated with antebellum horticultural introductions and specimen plants in formal gardens, including such botanical pests as tree of heaven and royal paulownias.

Home Gardens and Floral Culture

A few months into the Civil War in April 1861, the *Southern Cultivator* reminded readers, "It costs but little to have a neat flower garden, and to surround your dwelling with those simple beauties which delight the eye far more than expensive objects. Nature delights in beauty.... Follow her example, and do for yourself what she is always laboring to do for you."[20] Indeed most Americans had comparatively modest gardens that evolved with family needs and the occasional addition of new ornamental plants. Only a small percentage of the southern population owned large plantations with extensive gardens; by 1860 there were fewer than 50,000 plantations of 1,000 or more acres, and many of these were strictly agricultural and lacked a grand house and cultivated grounds.

Typical household horticulture included both kitchen and flower gardens, and good soil was essential for both edible and ornamental crops. Published when secession seemed likely, *Gardening for the South* (1856) devoted thirty pages to the soil of home gardens, including the need for crop rotation and the addition of manure, plant matter, guano, and night soil to improve soil nutrients.[21] Unlike fields and farmland that were often abandoned when nutrients diminished and crops waned, the land around homes required careful planning, planting, and management. *Gardening for the South* provided strictly practical advice, with sections devoted to vegetable crops, fruit production, and medicinal herbs, a move toward independence for Southerners who were able to garden. Some practices were remarkably forward-thinking, including the use of mulch, which "prevents moisture from evaporating, and it also prevents frost from penetrating to the roots. Mulching should also be applied to those herbaceous plants that are impatient of heat about the roots.... English peas are thus kept much longer in bearing, and rhubarb and other plants, requiring a cool soil, can be more readily raised.... Mulching not only wards off drought, but, in this way, by keeping the ground moist, and by the decay of the mulching substance, a good deal of food is conveyed to the plants. Some authors, indeed, think also that ground will become continually richer by being shaded."[22]

A half-acre wartime garden of legumes, cabbage, corn, sweet potatoes, and other root crops was sufficient for household survival even if farm crops were abandoned or plundered, although in 1862 the *Southern Cultivator* recommended that kitchen gardens also include tomatoes, eggplant, radishes, lettuce, and melons.[23] Seed-saving was encouraged, although the Confederate Ordnance Bureau imported large quantities of seed stock from Bermuda.[24] For many Southerners, successful kitchen gardens meant survival; by early 1864, the *Southern Cultivator* urged gardeners to "put in all the provision crops you can possibly work, and cultivate in the most thorough style, this year, as food with be scarce and high!"[25] In Washington, the Commissioner of Agriculture also encouraged farmers in Union states to plant gardens for their own families' needs.[26] Ideally these were established near homesteads and apart from agricultural fields; reading between the lines, women could care for dooryard kitchen gardens even if men were called away to military service. Suggestions for household crops were practical and included the use of hotbeds to lengthen the northern growing season by starting seedlings earlier in a warm environment.[27] Hotbeds were excavated, enclosed with glass (typically old window sash), and re-filled with layers of manure and soil. Heat from the metabolic activity of bacteria was sufficient to promote seed germination and protect against early spring frosts.

Even in wartime, celery (*Apium graveolens*) persisted as a mid-nineteenth-century dining fad, and many gardeners attempted to grow their own crops of this botanical oddity. Unlike other leafy greens, the edible parts are enlarged leaf stalks (petioles); each bears a tuft of small compound leaves and has pungent essential oils and outer strands of flexible support tissue (collenchyma). Victorians prized the pale stalks displayed in tall stands that were sold for this purpose, and a celery arrangement on the dining table was considered a sign of prosperity.

Exposed to the light, the stalks developed chlorophyll and more pungent terpene-based flavor, so growers adopted the labor-intensive process of blanching to maintain the pale color and mild flavor of well-cultivated celery.[28] Southern gardeners who cultivated successful crops followed an arduous protocol of "earthing up" soil around developing stalks to shield them from sunlight; gardeners in the North achieved the same effect by cultivating celery in trenches in which soil could be banked around the growing plants.[29] Selection eventually resulted in self-blanching varieties, as well as cultivars with crimson and golden petioles and unique flavors like 'White Walnut.'

Facing medical shortages, southern gardeners were advised to cultivate medicinal plants to supply household needs. Most of the recommended plants were familiar European herbs that were easily cultivated among flowers and vegetables in dooryard gardens. Most could be safely administered without fear of toxicity, and many are now known to have diverse antibiotic properties. Most dooryard herbs were weedy aromatic perennials that were used to treat common ailments such as colds, sore throats, colic, nausea, nervous complaints, and flatulence; several including dill (*Anethum graveolens*), anise (*Pimpinella anisum*), rosemary (*Rosmarinus officinalis*), and sage (*Salvia officinalis*) also had kitchen uses. Many early garden flowers were first grown in the New World as medicinal and household herbs, and although they were useful herbs, tansy (*Tanacetum vulgare*) and lavender (*Lavandula angustifolia*) were also often planted for their ornamental value in perennial borders. However, self-sufficiency determined the herbs needed in wartime, and plants with medicinal value were added to home gardens. A few years before the war, *Gardening for the South* recommended planting sesame for treating dysentery and boneset for treating severe fevers (see Chapter 5), although both were relatively new additions to the herbal pharmacopeia.[30]

Some garden herbs may have been particularly useful in wartime, including pennyroyal (*Mentha pulegium*), cultivated as an emmenagogue and used as an abortifacient.[31] Lemon balm (*Melissa officinalis*) was used to treat nervous conditions, and rue (*Ruta graveolens*) was an herbal treatment for hysteria.[32] Some plants were cultivated specifically as scurvy preventatives, including the European mustard known as scurvy-grass (*Cochlearia officinalis*).[33] An herbal extract of scurvy-grass was a standard remedy on sailing ships before citrus fruit were widely available. Other antiscorbutic garden plants included horseradish (*Armoracia rusticana*) and sorrel (*Rumex acetosa*).[34]

Flower gardening was a nineteenth-century pastime recommended for its moral and educational value. Most homes had vernacular gardens planted over time with seeds, cuttings, and offsets shared by others, and many garden plants were cultivated because they reseeded readily or self-cloned by roots or rhizomes. There usually was no pre-determined plan, so vernacular gardens evolved over time depending on available plants, personal preference, and horticultural fashion. Sometimes described as parlor gardens, the beds and borders of annuals, perennials, and shrubs were often planted with an intentional sight line from inside the home; growth was tangled or tidy, depending on plant selection and cultivation.

Home gardens were thought to reflect culture and refinement. Even as the Confederacy advised women to plan vegetable gardens for survival, the *Southern Cultivator* extolled flower gardening for its moral and educational value particularly in wartime, including a five-verse poem that reminded readers to

> Make your home beautiful—bring it to flowers;
> Plant them around you to bud and to bloom,
> Let them give life to your loneliest hours,
> Let them bring light to enliven your gloom.[35]

The extent to which women followed such suggestions varied, but in *A Journey in the Back Country* (1860), Frederick Law Olmsted complained that in most southern states he encountered domestic conditions so crude "I found no garden, no flowers, no fruit, no tea, no cream, no sugar, no bread."[36] He saw a stark contrast between the subsistence eked out by most Southerners and the large house, lawns, and statue-decorated gardens of a Natchez plantation.[37] Even modest homes could have admirable flower gardens, a central message of the *Ladies' Southern Florist* (1860), the only book published prior to the Civil War that dealt specifically with flower gardens for "every lady of education in the South."[38] Mary Rion included detailed information on growth habit, leaf types, and flower forms, reflecting the mid-nineteenth-century interest among women in botanical studies. As a young woman, she had studied botany with William Preston, the president of South Carolina College, and she integrated her experience as a gardener in South Carolina with information from standard works such as *The Flower-Garden* by Joseph Breck (1851) for gardeners in northern climates.

Mary Rion eschewed geometric beds and favored plantings that were round, oval, or irregular in outline. Boxwood hedges were optional, as was perfect symmetry, and she encouraged gardeners to plant roses and other shrubs amidst herbaceous plants. Through trial and error, she learned that the addition of acidity with "decayed leaves and swamp earth" caused the anthocyanins in hydrangea petals to become blue.[39] Rion's list of flowers suitable for southern gardens included species that tolerated heat, drought, and bright sun; that were easily grown from seed, cuttings, or roots; that produced copious viable seed for sharing, and that provided bright color and unique form. She ignored geography in favor of horticultural value and selected garden plants based on their ability to flower at some point in the growing season. She advocated transplanting native plants into garden beds, including yellow jessamine (*Gelsemium sempervirens*) and wild azaleas (*Rhododendron prunifolium* and related spp.).[40]

Many roses posed a horticultural conundrum after centuries of selection and hybridization, with some untraceable to native wild species in Asia, Europe, northwestern Africa, and North America. Southern garden roses included the climbing evergreen Banksia roses (*Rosa banksiae*), a Chinese species originally named for shrubs cultivated in the garden of Sir Joseph Banks. Tea roses also originated from China as varieties and hybrids of *Rosa odorata*, known for having a scent resembling tea. Other favored southern varieties were broadly categorized as bourbon roses, China roses, and hybrid perpetual roses.[41] China roses (*R. chinensis* and its hybrids) included the green rose (*R. chinensis* var. *viridiflora*), a garden curiosity selected from a mutant with a double calyx and no petals. Grafting onto stronger root stocks was common, accomplished by inserting buds into woody stems where they then grew into mature stems of the preferred variety.

Other recommendations from the *Ladies' Southern Florist* included old European garden species such as heartsease (*Viola tricolor*) and sweet William (*Dianthus barbatus*); these were colonial garden introductions that succumbed to intense summer heat, but they could be sown and cultivated in early spring flower beds.[42] However, Rion also followed some garden fads, including the cultivation of tropical bedding plants that became particularly popular after the Civil War. Such plants flowered continually in a hot climate, and their bright colors were eye-catching if not always in good taste. One such curiosity was cockscomb (*Celosia argentea* var. *cristata*), a garish mutation selected for the velvety floral mass that replaced normal flowers. Rion recommended sowing the crimson varieties, fertilizing with fresh horse manure, and removing side branches to produce "gigantic and magnificent combs" at the crown of each plant.[43] However, some disagreed with this method of soil enrichment; a columnist for the *Southern Cultivator* observed, "We have seen beautiful flower beds much disfigured by coarse stable manure spread upon the surface, and the perfume of the flowers did not always conceal a less agreeable odor.... A mixture of leaf mould, earth, and bone sawings is, on the whole, preferable."[44]

Additional advice for southern gardeners centered on seed-saving for garden continuity, although it is unknown how many women were able to continue flower cultivation during war.[45] Bulbs needed planting and lifting, dahlias demanded staking, and roses required pruning—all at a time when many Northern and Southern women were hoeing vegetable gardens and holding together their family farms. In Washington, the Department of Agriculture fostered a sentimental wartime view of ornamental gardens, encouraging flower cultivation for the sake of beautification, religious faith, and high cultural ideals, concluding, "Cultivated flowers are evidences of high civilization; they are a sort of floral thermometer, indicating in some degree the intelligence and refinement of the people; and their indications are as significant as are the evidences afforded by architecture, painting, poetry, or any of the sciences. The lessons of these gentle teachers are having an influence on the habits and manners of our people."[46] Even a cash-strapped householder could garden well, a window into his decency and work ethic. The 1863 *Report of the Commissioner of Agriculture* described simple dwellings surrounded by well-tended flower beds "amidst an air of tidiness, thrift, and comfort, and better farming generally."[47] Vines seemed to be a particular favorite; they could be trained to grow on fences or over doorways, producing an evocative image in the midst of wartime strife. Wild vines were easily transplanted to gardens, and these included native species such as virgin's bower (*Clematis virginiana*), Michigan rose (*Rosa setigera*), and Virginia creeper (*Parthenocissus quinquefolia*). Chinese twining honeysuckle (*Lonicera japonica*), an Asian introduction, was favored for its scented cream and yellow flowers, but it colonized aggressively by rhizomes and became invasive across some regions.

Wartime health was thought to be affected by flowers, including the "deep sorrows occasioned by bereavement.... The care of flowers will soften the poignancy of grief, and heal sorrows that no medicaments of the apothecary can cure; God's sunlight and pure air, the fragrance of unfolding flowers, the freshness of the soil, and free exercise among growing plants in the garden, may give health and life when all other means fail."[48] Some embraced garden work in particular as excellent for women's health generally. In *Plain and Pleasant Talk about Fruits, Flowers, and Farming*, Henry Ward Beecher advocated physical work and fresh air as antidotes to hours spent indoors with "embroidery, lace-work, painting rice paper, casting wax-flowers so ingeniously that no mortal can tell what is meant [by] lilies looking like huge goblets, dahlias resembling a battered cabbage."[49]

Indeed many women did spend hours doing needlework and fancywork, often replicating flowers and other botanical structures with thread, paper, wax, feathers, shells, ribbon, and human hair. Beecher argued that a fashionable life spent indoors led to morbid tastes, nervous disorders, and enfeeblement; in contrast, women who cultivated a love of flowers arose early, enjoyed good health, exhibited industrious habits, and developed powers of close observation. There was an element of truth to Beecher's thoughts because antebellum fashions and social expectations discouraged women from physical activity. Many no doubt suffered from vitamin D deficiency and weakness caused by inactivity, poor health potentially remedied by time spent gardening in the sunlight. A similar philosophy was advocated by a writer in the *Southern Cultivator*, who advised "to obtain rosy cheeks—Cultivate a flower garden. Rise early and try to discover where the roses and carnations get their brilliant complexions. It is a secret worth knowing. You will find the cosmetic you need in the same place."[50]

Indoor Gardens and Parlor Botany

Prior to the 1850s, tropical plants grew in the hothouses and conservatories of wealthy Americans, but houseplants were virtually unknown. Some believed that plants might deplete oxygen and replace it with carbon dioxide, with one common text warning that "house-plants, and bouquets in sick-rooms, are injurious; their influence on the atmosphere of the rooms is unhealthy."[51] However, interest in potted plants burgeoned as middleclass leisure time increased. Women who lived in towns and cities sought to beautify their homes, and they had the time and resources to pursue indoor horticulture. Interest in houseplants exploded in the decade prior to the Civil War, further encouraged by book publishers who released numerous titles on parlor gardening, window gardens, houseplant culture, and botany for interested amateurs.

The first American book on houseplants was *The Parlor Gardener: A Treatise on the House Culture of Ornamental Plants* (1861), which Cornelia Randolph translated and adapted from a French text.[52] She encouraged experimentation and propagation and provided general information about plant care, writing with a maternal tone: "Do not water them during their sleep, except when they appear evidently to be suffering from thirst; and then give them only as much as is necessary to relieve their suffering."[53] Her definition of parlor gardening included hanging plants, tabletop greenhouses, and aquaria for cultivating aquatic plants. Of particular interest were the various ornamental etageres and plant stands made of iron, wire, or wood on which several diverse plants could be arranged for admiration. Some were fitted with castors that allowed the stand to be relocated depending on household climate and light conditions. For those without funds for indoor botanical displays, the U.S. Commissioner of Agriculture suggested a cheaper, homemade alternative fashioned from twigs and branches: "Nothing is much prettier for the roomy parlor of a farm-house, where the surroundings are not too pretentious to harmonize, than a stand of light rustic work covered with thrifty, well grown plants, fringed with growing moss or fern."[54]

A particularly informative horticultural text was *Flowers for the Parlor and Garden* (1863) authored by Edward Sprague Rand, a Harvard-trained attorney who devoted his spare time to plant cultivation.[55] By 1876, *Flowers for the Parlor and Garden* had been issued in 25 editions, revealing the extent of popular interest in rearing plants indoors. Rand provided expertise in hands-on horticulture, from hothouse camellias and orange trees to potted begonias, primroses, geraniums, and cacti suitable for parlor windowsills. He grappled with practical matters including soils, pruning, diseases, and insect pests, and kept abreast of the vast number of horticultural varieties sold for home culture.

Amateur horticulturalists were attracted to flowers with remarkable traits and brilliant colors, particularly plants that doubled as parlor curiosities. Hybrids of Chilean *Calceolaria* species (*C. crenatiflora*, *C. corymbosa*, *C. cana*, and others) were admired for their red, orange, and yellow flowers, each with variable markings and petals fused into a pouch that could trap insects. The potted plants flowered for months on a cool windowsill, and they became a favorite of botanical illustrators because of their variability. Calceolaria hybrid seeds were widely advertised, including in the 1863 catalogue of the William R. Prince botanic garden and nursery in Flushing, New York. Other windowsill favorites included lobelias, petunias, salvias, and nasturtium, all now grown more commonly in outdoor beds and borders.

Many houseplants of the Civil War era were actually tender shrubs that were selected for compact growth and spectacular flowers. An article in the *American Agriculturalist* extolled the beauty of "the ladies' ear-drop," as some called the pendant flowers of the genus *Fuchsia*, named for the sixteenth-century herbalist and physician Leonhart Fuchs.[56] South American species were hybridized and selected for seasonal growth, double flowers, and color. The potted hybrids had petal and sepal colors in various combinations of white, pink, crimson, and purple; they lived indefinitely and flowered continuously with minimal care. Winter varieties were trimmed into single woody stems with lateral branches from which the flowers hung, and they thrived indoors with warm days and cool nights without central heating. Summer varieties were kept outdoors in shaded or sheltered borders or porches. By 1862, there were twenty popular fuchsias for home gardeners with names such as 'Psyche,' 'Florence Nightingale,' 'Queen Victoria,' and 'Prince Albert.' Other familiar shrubs for indoor cultivation included citrus trees, hydrangeas, jasmines, and flowering maples (*Abutilon pictum* and related hybrids in the family Malvaceae), which are mallow relatives and unrelated to true maples.

Pteridomania, a term coined for the Victorian love of ferns, began in England during the 1850s, and by the 1860s Americans cultivated specimens indoors and outdoors, in pots, urns, baskets, conservatories, ferneries, and landscaped rockeries. Interior rooms and hallways were often dimly lit, so shade-loving ferns thrived in pots and terraria; in gardens, ferns colonized deep shade where most flowering plants failed. Fern importers and dealers multiplied to meet demand, and fern societies and publications provided information for amateurs interested in fern culture. In 1861, J.L. Russell struggled to explain fern reproduction in an article for *The Magazine of Horticulture*, noting the dust-like spores were incorrectly advertised by nurserymen as "seeds."[57] He suggested germinating spores on a bed of sand and peat covered with a bell glass, but also noted that mature wild ferns could be easily transplanted to shaded gardens without the trouble of starting plants from spores. Indoor ferns such as maidenhair (*Adiantum* spp.) were often grown in mixtures of peat, sand, and cocoa hulls (the outer coats of cacao seeds, *Theobroma cacao*), and serious enthusiasts became specialists in coaxing growth from costly tropical specimens. For those who could afford the luxury of a separate room, indoor ferneries were lined with wire netting for climbing ferns, and the use of double pots kept soil moist.[58]

Botanically speaking, ferns differ from flowering plants in having two independent generations. Each microscopic fern spore germinates to produce a small green prothallus (often about a half-inch in size) on which fertilization occurs, resulting in the embryo which grows into a mature fern plant. Fern sperm require water, a process that occurred readily inside glass-enclosed terraria. Thus the problem of rearing and shipping ferns was solved by Nathaniel Bagshaw Ward, a London physician and naturalist who invented a sealed glass case for transporting imported ferns. Ward's booklet "On the Growth of Plants in Closely Glazed Cases" resulted in widespread interest in using the cases for ferns and other species that needed high humidity and protection from coal gases and other air pollutants. These small greenhouses became popular household furnishings in the 1860s, and various sizes of Wardian cases were made to use on tabletops or as freestanding terraria. They were planted with tropical or temperate ferns and other plants, depending on budget and interest. As an alternative, household manuals suggested inexpensive homemade cases assembled from glass panes and landscaped with local ferns, club mosses, and low-growing wildflowers such as partridgeberry.[59]

Houseplants and indoor gardening became a metaphor for family life and maternal devotion, particularly during the Civil War years when finances were tight, tensions high, and families separated. Encouraging worthwhile pastimes, the Commissioner of Agriculture encouraged families in Union states to raise houseplants, noting that "they adorn the house as nothing else can, and give to the cheapest furniture an air of elegance which no other ornament can impart ... they are a most graceful form of charity to the poor passer-by who has no means of gratifying his taste for the beautiful.... It is impossible to overestimate the effect of youthful association and daily companionship with such exquisite shape and coloring, which foster in the minds of children a taste for simple and natural forms of amusement and recreation."[60] *Godey's Ladies Book and Magazine* published numerous articles on horticulture, indoor gardening, and botanical topics for their wartime readership. In 1862, these included an article on forcing flowering bulbs in water, a report on a horticultural exhibit, and miscellaneous articles on color selection in gardens, plants in bedrooms, floral designs in fashion, and patterns for floral embroidery.[61]

Subsequent wartime volumes of *Godey's Ladies Book and Magazine* included instructions for cultivating flower gardens, growing plants from seeds, making artificial flowers from household materials, crafting rustic home decor, and poetry with botanical and floral imagery. An advertisement for *The Phantom Bouquet* (1863) directed *Godey's* readers to the new craft of skeletonizing plants, a process described in detail by Edward Parrish, a pharmacist and first president of Swarthmore College.[62] Parrish's method involved bacterial decomposition or boiling to macerate the soft tissue of leaves and fruit, bleaching in chloride of lime (calcium hypochlorite) or chloride of soda (sodium hypochlorite), followed by three-dimensional mounting. The text included a detailed appendix of plant parts recommended for skeletonizing, including deciduous and evergreen leaves, ivy and other vines, various fruits, and hydrangea inflorescences. Ghostly bouquets of pigment-free plants were an ideal Victorian decoration, but it is unknown how many women undertook the arduous process of their preparation.

In the North, Catharine Beecher and her sister Harriet Beecher Stowe viewed family parlors as a center of thrifty horticulture, practical botany, and craftwork using plant materials.[63] Their suggestions included picture frames made from twigs, pine cones, and acorns; woodland baskets; pots of fuchsia, trained ivy, and other trailing plants; hanging pots made from coconut shells; a flower stand made of tree roots; and a homemade Wardian case filled with native ferns and other woodland plants. They described the value of a Wardian case as an educational indoor garden and encouraged children to keep busy with plant-collecting and gardening.[64] J.H. Walden, author of the standard farming text *Soil Culture*, agreed, noting, "Association with flowers should be a part of every child's education. Their cultivation is suitable for children and young ladies in all the walks of life."[65]

Not to be confused with Wardian cases, Waltonian cases were invented as miniature greenhouses designed to encourage root growth by maintaining propagated plants at an ideal temperature. Each case consisted of a zinc tray enclosed by glass panels and warmed with an oil lamp. In homes without central heat, they served as tabletop greenhouses to protect pots of expensive tropical plants on chilly winter nights. Waltonian cases particularly appealed to women who did not want to carry prized houseplants outdoors to hotbeds heated by a layer of manure (see Chapter 4). Regarding the convenience of Waltonian cases, an article in the *American Cotton Planter and Soil* noted, "Such a little hothouse can, like another piece of furniture, be placed near the window in any room and will answer perfectly for forcing or wintering all tender plants."[66]

Home decoration during the Civil War years often included botanical art and floral designs, and one such illustration hid a political message. Following the 1862 capture and occupation of New Orleans by federal troops, many homes displayed the "Fleurs du Sud" lithograph by J.B. Guibet. The seemingly innocuous floral design in fact represented the "stars and bars" of the first Confederate flag; Guibet used red and white flowers to delineate red and white stripes and blue and five-petaled white flowers to represent the canton. The flowers are fairly generic in style (probably poppies, petunias, carnations, and composites), but this was political rather than botanical art, and Guibet's intent was not floral accuracy. When Union General Benjamin Butler discovered the symbolic significance of the popular lithographs, he attempted unsuccessfully to have both the lithographs and original engraving plates seized and destroyed.

By the 1860s, the romanticized idea of a language of flowers diminished after a few decades as a popular amusement. During the first half of the nineteenth century, numerous volumes suggested and interpreted the symbolic meaning of flowers, although often inconsistently.[67] Otherwise serious women took up the fad. Dorothea Dix, who served during the Civil War as the superintendent of Union army nurses, authored the *Garland of Flora* (1829), which integrated floral meanings with lines from poetry and other literature.[68] Sarah Josepha Hale, editor of *Godey's Lady's Book and Magazine*, wrote *Flora's Interpreter: or, the American Book of Flowers and Sentiments* (1832), which remained in print through the Civil War years.[69] Even standard textbooks like *Botany for Beginners: An Introduction to Mrs. Lincoln's Lectures on Botany* (1833) devoted a few pages to familiar flowers and the sentiments that they conveyed.[70] Some meanings were predictable: Lilies denoted purity, and red roses suggested romantic love; other meanings were less obvious, including the symbolism of iris as a "message for you." By the Civil War, the notion of a floral language to communicate romantic messages had lost much of its early attention, although many of the books still remained in American parlors.

Gardens and Slavery

Enslaved people often cultivated vegetable gardens near their cabins, in plots of varying sizes that may have included ornamental plants. Yards and gardens around cabins were strictly vernacular, and their intended purpose was to fulfill the daily needs of the gardeners who tended them. Some enslaved people also gained a degree of independence by gardening; this was certainly the case at Monticello, where enslaved gardeners grew fruit and vegetable crops which they bartered with the Jefferson household as a means to earn some independent income. Recorded transactions reveal that enslaved people grew garden crops in season, carefully stored the produce in cool root cellars, and then sold fresh fruits and vegetables months later for use in the Jefferson household.[71]

As observed by Frederick Law Olmsted in his travels through southern states, food was the primary need, and enslaved people were encouraged to cultivate vegetables for their tables (see Chapter 4). Gardening was often done by moonlight or firelight because days were occupied by plantation work. Agricultural journals reminded slaveholders that it was in their best economic interest to encourage enslaved people to garden successfully to provide "wholesome and nutritious vegetables for their negroes, such as cabbages, potatoes, turnips, beets, peas, beans, pumpkins, & c."[72] Some planters believed that gardens encouraged pride, making it less likely that workers would try to escape; others forbid individual plots because they feared independence among enslaved people and instead designated central gardens for communal use. Cash crops were also a matter of owners' preference. Some owners were liberal about allowing enslaved workers to cultivate and sell cash crops such as corn and cotton raised on their own time, while others forbid the practice. Gardens may have been particularly important for enslaved women, who ran the risk of having their husbands sold. Cash crops grown on a few or several acres would have afforded them a degree of independence and the ability to provide for their children.[73]

Gardens near cabins typically comprised beds surrounding a swept yard, often a shared area in which the earth was compacted by use and kept tidy by routine weeding and sweeping. Water ran off the hardened clay layer which resembled pavement, and so swept yards helped to control sticky gumbo mud during rainy seasons, apparently a tradition introduced from western Africa.[74] Daily sweeping was done with homemade brooms, crafted from the twigs of various shrubs, including gallberry (*Ilex glabra*), dog fennel (*Eupatorium capillifolium*), and dogwood (*Cornus florida*).[75] The WPA narratives included descriptions of swept yards, but the practice was not limited to those with African heritage.[76] In short, swept yards became a rural rather than racial tradition. Some white families also compacted and swept their yards out of practicality, but there were critics of the practice. Landscape gardener George Kidd advised against it, suggesting "let all the ground around the dwelling be kept loose and cultivated … the practice of sweeping all litter from the trees and much of the surface soil with it rendering the soil impervious to the action of the atmosphere are the principal causes of the trees declining."[77]

The preferred trees were crepe myrtles, which recalled the myrtle (*Myrtus communis*) mentioned in the Old Testament, although they are not closely related species. However, aside from providing some shade, trees were not an essential element in the gardens cultivated by enslaved people in which plants were grown for either food or flowers. Gardens were an extension of kitchen and living quarters, areas where outdoor work and visiting occurred, and personal preference and resourcefulness factored into their design and arrangement. Pathways and beds were edged with found objects, including bricks, jugs, stones, and glass bottles of different sizes and colors. Following an African tradition, scavenged bottles were inserted over the trimmed branches of shrubs or small trees. The preferred color was cobalt blue, which recalled the indigo pigment (see chapters 3 and 7) used to tint house paint and purported to keep evil spirits away from homes. According to legend, the bottles captured evil spirits and held them until they were killed by sunlight.[78]

The WPA narratives document the planting and enjoyment of flower beds and the care of honeysuckle, crepe myrtle, climbing roses, and other flowering shrubs. Swept yards were often surrounded by beds of flowers and vegetables, sometimes with ornamental and food plants comingled. Some gardeners transplanted native shrubs and wildflowers, but it seems that most preferred tropical bedding plants such as petunias, cockscomb, zinnias, verbenas, and marigolds. These were the same bright-flowered cultivars grown in mid-century parterre gardens, easily cultivated in a warm climate and shared from saved seeds. Traditional medicinal herbs were also cultivated (see Chapter 5), including rosemary, tansy, sage, mullein, horseradish, horehound, catnip, yarrow, and peppermint. Native medicinal plants cultivated by enslaved people included coneflower (*Echinacea purpurea*, also known as Sampson's snakeroot), and butterfly weed.[79]

The WPA narratives mention the most common fruits and vegetables in cultivation, including muskmelon, watermelon, turnips, sweet potatoes, white (Irish) potatoes, peanuts, corn, peas, beans, and cabbage (see Chapter 4). These garden crops yielded essential vitamins and other nutrients and thus some historians have argued that gardening provided enslaved people with better diets than those of many white Southerners.[80] Using seeds probably introduced from Africa, enslaved people cultivated bottle gourds (*Lagenaria siceraria*); grown to maturity, the dried pepos (the fruit of the gourd family, Cucurbitaceae) were trimmed into bowls and water dippers, recalled by the lyrics of a traditional song. Enslaved people sang "Follow the Drinking Gourd" to share escape instructions among plantations by word of mouth. With the aid of the Underground Railroad, the journey on foot took about a year. The drinking gourd refers to Polaris, the large North Star at the end of the handle of the Little Dipper constellation; the two stars of the bowl of the Big Dipper

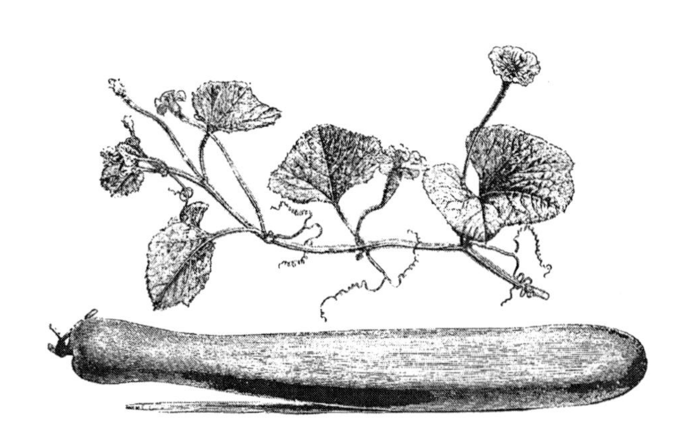

Enslaved people cultivated bottle gourds to make into bowls and water dippers, recalled in the song "Follow the Drinking Gourd" which shared escape instructions for using the Underground Railroad (morphart © 123rf.com).

constellation point toward Polaris.[81] The last verse of the song refers to the Tennessee River and Ohio River, where escaped workers crossed the frozen river to meet their guides on the north bank:

> When the great big river meets the little river,
> Follow the Drinking Gourd.
> For the old man is waiting to carry you to freedom
> If you follow the Drinking Gourd.

The nutritional importance of these gardens continued after the Civil War, and for many former owners and overseers, having emancipated enslaved people remain on plantations was economic good fortune. Lacking labor, landowners were unable to keep fields clear and cultivated, and pine forests soon colonized land that had once provided food for plantation homesteads. However, freedmen cultivated crops and bartered food with former slaveholders. During the Depression years of the 1930s, seeds were sometimes difficult to obtain, but older formerly enslaved people saved seeds after each growing season and continued to cultivate subsistence gardens.

Botanical Studies

Botany is a pure science with connections to agriculture and medicine, and practical knowledge of plants enabled wartime progress in both disciplines (see chapters 3 and 5). However, the process of documenting North American flora was still a work in progress, and during the antebellum years, botanical classification kept pace with geographic exploration. Professor Asa Gray at Harvard University trained as a medical doctor, but he soon turned his attention to botany and assisted John Torrey with sections of the *Flora of North America* (1838–1843). Gray then authored *A Manual of the Botany of the Northeastern United States*, a standard reference work that was published in eight editions during the next century.[82]

Gray closely observed the similarity between the flora of eastern North American and eastern Asia, a phenomenon that many botanists later referred to as the Asa Gray disjunction because of the geographic "gaps" between closely related plants. In particular, Gray was curious about the relationship of North American plants to their Asian counterparts, some of which horticulturalists and gardeners were propagating and cultivating widely in America (see Botanical Origins above). Gray corresponded frequently with Charles Darwin on diverse topics, and he became the foremost American defender of Darwin's theory of evolution by natural selection, which some viewed as heresy. During the 1850s, Gray and Darwin discussed climate and other factors that might contribute to the distribution of similar species in areas as distant as New England and Japan. During the same decade, Japanese specimens arrived at Harvard as part of the U.S. Surveying Expedition to the North Pacific Ocean; Gray noted their striking similarity to closely related species in North America.

As biologists, they discussed slavery. In a letter to Asa Gray in which he also described the pollination of primroses and orchids, Darwin observed that sympathy in England was with the North and noted, "Great God how I should like to see that greatest curse on Earth Slavery abolished."[83] He had detested slavery since he witnessed the beating of an enslaved child in Brazil (see Chapter 1). Well into the war, Gray wrote to Darwin with news on his recent papers on *Astragalus*; he had described just one species of milk vetch in the first edition of *A Manual of the Botany of the Northeastern United States*, but numerous new species were discovered in the western U.S. He reminded him, "The sentiment of our country, you must see—at least I assure you—has settled—as I knew it would if the rebellion was obstinate enough—into a determination to do away with Slavery."[84] Following the end of the war and the Thirteenth Amendment to the United States Constitution, Darwin wrote to Gray about pollination mechanisms and recent revisions to *Origin of Species*, closing with "I declare I can hardly yet realise the grand, magnificent fact that Slavery is at end in your country."[85]

Southern botanists were also at work during the antebellum years. Some, like Asa Gray, began as physicians; most medicines at the time were derived from plants, and indeed for many years botanical expertise was essential to medical practice. Charles Wilkins Short practiced medicine in Kentucky and taught *materia medica* as a professor of medical botany. Short botanized extensively along the Ohio River and in 1833 published *A Catalog of the Native Phaenogamous Plants and Ferns of Kentucky*. He discovered several new species, including a goldenrod (*Solidago shortii*) and a sedge (*Carex shortiana*). Asa Gray named the genus *Shortia* to commemorate Short, with whom he corresponded. The genus of rare wildflowers includes five species (four from Asia and one native to the Appalachian Mountains) and is a typical example of the Asian-North American disjunction that Gray observed.

Edmund Kirby Smith graduated from West Point and served as a Confederate general, but his expertise lay in mathematics and botany. In 1854 he was botanist in the Mexican-American Boundary Survey, where he and his colleagues collected thousands of herbarium specimens along the border with Mexico. He was the last Confederate general to surrender at the end of the Civil War, and ten years later became a professor of mathematics and botany at the University of the South in Sewanee, Tennessee. Smith often botanized with Augustin Gattinger, a German physician and botanist who settled in Tennessee in 1849. He practiced medicine as a country doctor and collected widely at a time when much of the local flora was relatively undocumented. After serving as a surgeon with the Union army, Gattinger used his medical expertise and botanical collections as the basis of two volumes, *Medicinal Plants of Tennessee* (1894) and *Flora of Tennessee and Philosophy of Botany* (1901).

Botany was the first science taught to young students, and it was often used to interpret the natural world through a religious lens. Although a believer in evolutionary theory, Asa Gray published botanical textbooks that encouraged children to see God's dominion on earth; the first chapter of *How Plants Grow* began with a verse from the Sermon on the Mount (Matthew 6:28, 29), "Consider the lilies of the field, how they grow; they toil not, neither do they spin; and yet I say unto you, that Solomon in all his glory was not arrayed as one of these."[86] To introduce botanical diversity and classification, he reminded children that "God created plants in a vast number of kinds." and grouped related plants in families.[87]

Before the Civil War, Gray had abandoned the eighteenth-century Linnaean system of classification that he used in *Elements of Botany* (1836). In his *Systema Sexuale*, Linnaeus defined taxonomic classes based on stamen number and orders based on pistil number; thus mountain laurel (*Kalmia latifolia*) with ten stamens and one pistil was classified with the *Decandria Monogynia*, while milkweed (*Asclepias*) with five stamens and two pistils was assigned to the *Pentandria Digynia*. Almira Hart Lincoln Phelps had used the Linnaean system in *Botany for Beginners: An Introduction to Mrs. Lincoln's Lectures on Botany* (1833), which encouraged students to find, identify, press, mount, and label specimens for their own herbarium collections. This was likely the first hands-on laboratory work in American classrooms, a significant departure from the standard curriculum of drill and recitation. Although ministers and mothers raised concerns about references to sexual function, the Linnaean system was widely studied during the antebellum years, particularly by young women. *Botany for Beginners* identified Linnaean classes and orders using stamen and pistil count, but avoided the term *Systema Sexuale*. In *Peter Parley's Illustrations of the Vegetable Kingdom* (1840), Samuel Griswold Goodrich explained "the Linnaean or function of botanical 'male and female organs' as analogous to human structures, which may have shocked some readers."[88] The title page illustrated children in a garden, a message supported by social reformers who encouraged youthful interest in botany and horticulture. Goodrich was also progressive in his discussion of soil, latitude, and altitude and their effects on plant growth, an early ecological concept.

Of particular interest in teaching children were plants with peculiar traits, such as the sensitive plant (*Mimosa pudica*) which responds to touch and scarlet pimpernel (*Anagallis arvensis*), which Goodrich described as "one of Flora's timepieces, opening its flowers about eight o'clock and closing them about two."[89] Pitcher plants (*Sarracenia purpurea*), one of the most curious American species, were still poorly understood, although Asa Gray and the 1862 *Report of the Commissioner of Agriculture* both described them as unique bog or swamp plants.[90] However, neither observed that its unique leaves trap insects as a source of nutrients, a fact that was not realized until the 1880s.

7

Fibers and Dyes

King Cotton and Culture

Cloth, cordage, and paper depended on cotton, flax, and hemp, the fiber crops grown on U.S. farms and plantations. Secessionists used "King Cotton" as a slogan for an independent Confederate economy with diplomacy based on cotton trade, without which the New England textile mills would fail. King Cotton advocates began by burning 2.5 million cotton bales to create a fiber shortage, followed by halting cotton shipments north in early 1861. Southern cotton was the leading antebellum export to Europe and provided the raw material for textile mills during the Industrial Revolution. Secessionists thus used cotton in an unsuccessful attempt to leverage British and French support for the Confederacy, while Northern agents continued to buy cotton for use in Union states.

Despite government advice to grow food rather than cash crops, the planters on many large plantations continued to cultivate wartime cotton crops. In some cases, overseers wanted to keep field workers busy because they feared violent uprisings. Enslaved people were also tasked with guarding the cotton bales hidden in houses, barns, cellars, and swamps, awaiting buyers from the North. Cotton did supply some Confederate military needs, ranging from tents and uniforms to bandages and a quinine substitute from cottonseed (see Chapter 5); its cultivation was seen as an act of defiance against the federal government. Bales were taken to pay the 10 percent in-kind tax on crops (see Chapter 3), but in anticipation of a Confederate victory, much of the cotton grown during the war years was stored to sell after the war. Once emancipated, many freedmen remained on plantations seized by the Union, where they sorted cotton that was sold for revenue or used for manufacturing needs.

The Confederate economy largely pivoted on cotton, and during the war, the Confederacy bartered cotton for armaments, drugs, and other necessities from England. Ships carried cotton to British ports in Nassau and Bermuda and returned to southern ports with loads of essential goods, including coffee and bluestone, the dolomitic mineral that was thought to prevent fungal diseases in wheat seedlings. As a result of this trade, Southern planters provided over three-fourths of the cotton spun and woven in British mills. Cotton paid for Confederate warships that fought the Northern merchant marine, and it was only the capture of southern ports that quelled the bartering of cotton for imported supplies. Some planters sold their cotton directly to British agents, and they burned whatever could not be sold or bartered before the arrival of Union troops. Union soldiers confiscated any remaining cotton bales and shipped them north, where sales often resulted in personal profit to them or their families.

At the same time, cotton cultivation increased in Egypt and India with the opportunity to replace Confederate cotton in the world market. Egyptian varieties originated with Sea Island cotton (*Gossypium barbadense*), and crops became known for producing particularly long fibers. Indian farmers cultivated both Asian cotton (*G. arboreum* and *G. herbaceum*) and varieties derived from the New World species (*G. hirsutum* and *G. barbadense*) cultivated in southern states. Southern planters denied the possibility that American cotton could be replaced by foreign crops; in early 1862, the *Southern Cultivator* defiantly declared "there is no dependency upon India for Cotton, and that to the Cotton States of the Southern Confederacy England and all Europe must look for this great staple now; henceforth and forever."[1] Yet British concern for cotton shortages resulted in proposals for cotton alternatives. *The Illustrated London News* published a short article about the submerged aquatic plant known as grass wrack, sea wrack, or eelgrass (*Zostera marina*), accompanied by an engraving captioned "Mr. Harben's Proposed Substitute for Cotton."[2] Based on the strength of its fibers, an 1862 presentation to the British Association for the Advancement of Science extolled its potential as a cotton substitute. The stringy dried stems of eelgrass had long been used to stuff mattresses, thatch roofs, and insulate walls, but ultimately the plants proved useless for spinning thread and weaving cloth. As shortages of American cotton caused economic hardship in English mills, the *Bury Guardian* published a poem that satirized eelgrass as a cotton alternative, including these lines:

> Zostera Marina, grim Manchester's shaking,
> One half of her steam-engines silent and still,
> No cotton's at hand, and we're all in a taking
> To know where to turn, for new grist for the mill.[3]

Cotton culture was deeply ingrained in Confederate agricultural tradition, and southern growers developed the varieties cultivated during the Civil War. Most originated with upland cotton (*Gossypium hirsutum*), perennial shrubs native to Central America. When dry, each mature boll splits to reveal the fibers (see Chapter 3); each fiber begins as a rounded outgrowth from an epidermal cell, which then elongates and spirals into a trichome or plant hair. Typical bolls contain about ten seeds, each covered by up to 20,000 fibers, which vary in length between the shorter linters (about 25 mm) and the longer staples or lints (30–40 mm). These seed fibers were the starting point for the cotton industry, which progressed quite slowly until the invention of the cotton gin. The point of attachment between fibers and seeds is remarkably tenacious, a detail of plant anatomy that explains the importance of gins to the antebellum economy.

Prior to genetic selection, cotton was not entirely suited to cultivation in southern states. Wild shrubs were frost intolerant and flowered during late summer and early fall; botanically speaking, they are short day plants, species that require a certain minimum amount of darkness in order to flower. On a practical level, early planters soon realized that wild populations flowered too late in the summer for the bolls to mature and produce seed fibers during the southern growing season. With the expansion of the cotton industry, growers across the South sought varieties that suited regional climates and soil. During the antebellum years, natural variation in boll size and flowering time provided the genetic raw material for breeding and selecting improved upland cotton varieties.

Wild cotton from Mexico had particularly desirable traits that were introduced into the gene pool of cultivated cotton; the shrubs resisted disease, their bolls matured quickly, and the fibers were easily harvested. Development of new cotton varieties was accomplished with rudimentary knowledge of inheritance, and the process relied solely on the close observation of cotton crops in the field. Growers intuitively planted desirable varieties close-by, where hybridization occurred as a result of cross-pollination. By carefully saving and planting seeds from shrubs that produced the largest bolls in the summer, farmers developed cultivars that produced abundant fibers before the end of the growing season. The most promising hybrids were identified and protected from crossing with less desirable strains in order to yield seed stock for planting. The antebellum development of new cotton cultivars predated Gregor Mendel's experimental crosses of garden peas (1856–1863) which revealed the basic principles of genetics.

King Cotton became the most important southern cash crop before the Civil War. Cotton farmers developed new varieties by planting shrubs with desirable traits nearby each other for cross-pollination. Promising hybrids were identified and protected from crossing with less desirable strains in order to yield seed stock for planting (author's collection).

The most popular colonial cotton variety was 'Georgia Upland' (also known as 'Georgia Green Seed'), from which farmers selected plants which thrived in particular regions and yielded fibers suitable for spinning. A wild species (*Gossypium barbadense*) from the Caribbean and perhaps Brazil produced longer, finer hairs, but fiber yields were lower and harvesting was comparatively difficult. Cultivated varieties became known as Sea Island cotton, and their successful cultivation was limited to the coasts and coastal islands of Georgia, South Carolina, and Florida. By

1820, growers had developed several regional varieties of both upland and Sea Island cotton, including possible hybrids of the two species. Newspapers and agricultural journals printed numerous advertisements for cotton seed that promised excellent crops, including varieties such as 'Alverado,' 'Petit Gulf,' and 'White Nankeen.'

During the war years, at a time when growers were encouraged to plant cotton fields with corn (see Chapter 3), the *Southern Cultivator* continued to advertise 'Zipporah' cotton, which promised prolific growth and "market value DOUBLE, or more than double that of any other kind."[4] However, growers noticed that varieties often became "run down" over time. In short, crops became less productive and more susceptible to insects and diseases, physiological deterioration resulting from cross-pollination between new varieties and older, unimproved strains. In the 1850s, growers grappled with fungi and moth larvae (various "worms") that sometimes destroyed half of the harvest. Seed rot (see Chapter 3) caused decay of the bolls and fibers from the inside out, bollworms ate the developing fruit, and cotton worms reduced leaves to skeletons. Genetically improved varieties were often resistant to such microbes and pests.

New cotton cultivars proliferated in the 1840s and 1850s, including the 'One Hundred Seed' variety developed by Colonel Henry Vick of Mississippi. He tasked field workers with collecting the largest bolls from the most prolific shrubs and then planted the seeds in a common field for cross-pollination. From the progeny, Vick personally selected a single ideal boll from which he produced commercial seed stock, and he named the new variety based on the supposed number of seeds in each boll.[5] 'One Hundred Seed' became a favorite across all of the cotton-producing states because Vick sold seeds at a bargain price (typically $1.50 per bushel), and crops yielded good quality fibers. Of course, at a time when capitalism prevailed and there were few if any legal deterrents to the theft of cultivars, there was always the risk that seeds obtained from the original breeder could be used to produce seed stock for a "new" variety. Vick's neighbors soon marketed cultivars that they named 'Banana,' 'Pomegranate,' 'Hogan,' and 'Cluster,' which seemed to be genetically identical to 'One Hundred Seed.' Most successful was 'Jethro' cotton, sourced from seed shared by Vick and grown by Jethro Jones, the editor of *Southern Cultivator* who used the magazine to advertise seed stock to Georgia growers.[6]

Some varieties were local favorites but not widely cultivated. Planters tended to hold strong opinions about the varieties worth cultivating, probably based on the physiological "fit" between cultivars and local climates and soils. One such regional cotton was 'Mastodon,' which was valued by Mississippi Valley growers because its fibers remained intact despite wind and rain, a trait that allowed harvesting late in the season. Yet despite its impressive name, some planters regarded the variety as subpar and liable to deteriorate genetically over time.[7] Most important, however, was the overarching desire for efficiency and productivity, the combination of capitalist zeal and genetic selection that drove the southern cotton economy. By intentionally breeding and cultivating strains with high yield and quality fibers, planters quadrupled the rate at which enslaved people picked cotton during the sixty years prior to the Civil War.[8]

Cotton culture was a way of life, and during the Civil War there were frequent reports of innovation using unsold cotton crops. Before refrigeration, raw cotton fibers were used for food preservation; fruits such as apples, pears, and grapes were stored in glass jars between layers of cotton, and the lids were sealed with wax. It was thought that cotton somehow prevented ripening; before the role of microbes in decay was completely understood, cotton seemed to work "by arresting the germs or eggs of infusaries which are constantly floating in the air, and which are probably the approximate cause of all putrefaction."[9] In reality, the cotton layers probably allowed for the more rapid diffusion of ethylene, which could cause over-ripening and decay (see Chapter 4). A wartime substitute for the strong linen thread used in shoe manufacture was made by twisting eight strands of thinly spun cotton together into a thread suitable for lacing with a needle.[10] The flammable nature of cotton also made it ideal for lamp wicks, leading to a wartime invention that inserted cotton fibers into an outer casing for efficient burning: "The Excelsior wick ... is tubular, with raw cotton filling, having complete capillary attraction, making it a perfect feeder for all the heavier oils."[11]

Cotton Gins

Various attempts at mechanical ways to separate cotton fibers from seeds reflect the historical struggle to prepare fibers for spinning. In ancient India, textile workers removed the seeds by rolling fibers on flat stones using wooden or metal rods, a method that evolved into the manually operated machines or "engines" that are now known as cotton gins. By the twelfth century, devices with two rollers were widely used in India and China, soon followed by gins with worm gears and crank handles. The first cotton gins used in the American colonies were modelled on Indian technology and were the starting point for mechanical invention on a larger scale.

Each fiber is the outgrowth of a single cell on the seed coat, and with long staple cotton such as the cultivars of Sea Island cotton (*Gossypium barbadense*), rollers worked well in breaking the attachment point between fiber and seed. However, upland varieties (*G. hirsutum*) had shorter staples that resisted separation with early dual roller gins. The seeds were notoriously sticky and completely covered by hairs, which had to be removed by hand before the cotton could be spun. Two pounds of cotton required a day to process by hand, work that was done by enslaved people during rainy weather or evening hours.[12] Colonial inventors tinkered with dual roller machines that slowly squeezed the seeds from the mass of fibers, but processing remained economically inefficient until Eli Whitney devised a different mechanism.

Whitney envisioned his cotton gin design while working as a tutor on a Georgia plantation, but he relied on memories of textile machines in his native Massachusetts for mechanical inspiration. His handwritten patent application revealed insight into the potential of cotton as a cash crop: "green seed cotton, which was so extremely difficult to clean, as to discourage all further attempts to raise it—That it was generally believed that this species of cotton might be cultivated with great advantage if any cheap and expeditious method of separating it from its seeds could be discovered."[13] He responded with an invention that processed 50 pounds of fibers daily by using a wooden cylinder covered with wire teeth. As the cylinder rotated, cotton was processed as the rows of teeth passed through screen-like grids that captured the seeds but not the lint.[14] Whitney's ginning mechanism was vastly more efficient and thorough than removing seeds by hand.

Eli Whitney and his partner, Phineas Miller, built several cotton gins and installed them throughout the South, charging two-fifths of each crop ginned as an in-kind payment. However, many planters observed the machines at close hand and built their own versions to avoid future payments, which in effect released the design into the public domain. Although Whitney successfully patented his gin in 1793, issues soon followed with similar machines and dubious patents from competitors who made only minor modifications to the original design. Other types proliferated during the antebellum years, including the McCarthy gin that was designed specifically to process the long staples of Sea Island cotton which were damaged by Whitney gins. Planters in Georgia, South Carolina, and Florida adopted the McCarthy gin, which was a roller design augmented by reciprocating blades that released the seeds without damaging the long staples.[15]

By mechanically separating fibers from seeds, gins converted cotton into a highly profitable crop that was well suited to southern climates; processing time was no longer a limiting factor. Cotton harvests across the South nearly doubled each decade leading up to the war, and demand increased for field labor and cotton gins. By 1860 there were fifteen states that allowed slavery and 57 manufacturers of various mechanical gins, most of them minor modifications of Whitney's patent. One in three people in the South was enslaved, and two-thirds of them worked on cotton plantations. A vicious cycle evolved that pivoted on cotton fibers: Crops were needed to buy enslaved workers, and enslaved workers were needed to raise crops. The production of cotton bales increased from 3,000 bales in 1790 to 3.5 million by the 1850s, produced by the labor of millions of enslaved people.

Cotton gins used a rotating cylinder to draw cotton fibers through screen-like grids which captured the seeds but not the lint. The ginning mechanism was more efficient than removing seeds by hand, allowing the antebellum cotton industry to burgeon (*Harper's Weekly*, 1869).

Demand also increased for land, which was frequently taken from Native Americans and converted to cotton cultivation.[16] Geographically speaking, both slavery and agriculture shifted south, with nearly a million enslaved people being relocated from Virginia, North Carolina, and South Carolina to cotton plantations in the Deep South.[17]

Bast and Leaf Fibers

In the years leading up to the Civil War, cotton was the major American export. However, it was shunned worldwide by Abolitionists who followed the free-produce movement, an international boycott of slave-produced crops and other goods. In the United States, free-produce stores were operated by Quakers who offered for sale provisions and dry goods with no ties to enslaved labor. Flax was the primary alternative to cotton, but the linen cloth woven from flax fibers was less comfortable and more costly because supplies were often imported from Europe. Women with Abolitionist leanings substituted linen clothing for stylish cotton dresses; the residents of Fruitlands, the agrarian commune satirized by Louisa May Alcott in *Transcendental Wild Oats*, wore only linen clothing even during New England winters.[18] With a philosophy that also forbid the exploitation of animal products such as wool, Bronson Alcott and his followers subsisted largely on fruit, while shunning sugar and cotton due to their dependence on enslaved labor.

Union farmers grew flax for northern mills, but the bast fibers required chemical processing to resemble the texture of true cotton. The plants were also cultivated as a source of linseed oil used in making soap and waterproofing cloth (author's collection).

Flax (*Linum usitatissum*) is probably the oldest fiber used in making textiles. The plants are annual herbs native to Europe and Asia, but the species travelled worldwide with human migration, and no truly wild populations still exist. Puritans carried seeds to America where flax crops were cultivated for their bast fibers, flexible elongated cells which develop in the phloem (food conducting tissue) of the stems. By the mid–1800s, farmers in Union states grew flax for its seed oils, but its fibers were regarded as the most likely substitute for wartime shortages of southern cotton. Information from the Department of Agriculture noted that flax fiber production had dwindled between 1849 and 1859, and the agency aimed to re-educate farmers on routine matters of soil preparation and harvesting. Northern farmers were encouraged to grow wartime crops as a cotton alternative, and initially there was hope that flax could be mechanically "cottonized" so that its fibers would be more suited to spinning and weaving in New England textile mills. In reality the conversion of flax to "cotton-flax" involved a dangerous and complex chemical process. The treatment involved the immersion and boiling of the fibers in sulfuric acidic, sodium hydroxide, and sodium carbonate until the extremes of acidity and alkalinity loosened the cellulose fibrils of flax to resemble a cotton-like texture.[19]

Flax was also of interest in Confederate states where specific cultivars were selected for high oil yield in the seeds or long bast fibers for making linen cloth and paper. Fiber processing began with retting which relied on environmental microbes. Flax stalks were spread on the ground where they were exposed to dew, rain, and fungal spores, and after several days of exposure, the bast fibers were easily separated from the remaining stem tissues. During a time of shortages, flaxseed was used medicinally to brew a demulcent tea, and the linseed oil from pressed seeds had wartime uses for making soap and waterproofing cloth.[20]

Other wartime uses for flax included surgical lint, an essential medical supply used by both Union and Confederate armies for dressing wounds. The product consisted of either linen strips used as bandages or repurposed flax fibers from worn linen cloth used for their absorbency. The term *charpie* was used synonymously with lint, sometimes referring in particular to a type of lint in which narrow linen strips were partially unraveled into loose threads. A similar commercial product known as patent lint was made from linen cloth (or occasionally cotton) by selectively removing threads to produce a loosely woven bandage. Civilian war work included lint production; societies of volunteers organized to roll surgical bandages and scrape worn household linens into clumps of absorbent flax fibers. This loose lint was wrapped in muslin and applied directly to wound surfaces before bandaging. Typically the mass of fibers was moistened with water to prevent crusting or scabbing, which at the time was not regarded as part of the healing process; a covering of oiled silk (silk cloth treated with linseed oil) was used to prevent evaporation.[21]

When medical supplies ran low during the summer of 1862, Surgeon General William Hammond appealed to readers of the *New York Times* for help, "To the Loyal Women and Children of the United States: The supply of lint in the market is nearly exhausted. The brave men wounded in the defence [*sic*] of their country will soon be in want of it. I appeal to you to come to our aid in supplying us with this necessary article. There is scarcely a woman or child who cannot scrape lint, and there is no way in which their assistance can be more usefully given than in furnishing us the means to dress the wounds of those who fall in defence [sic] of their rights and their homes."[22] In her poem "The Women Who Went to the Field" (1892), Clara Barton mentioned the preparation of lint as typical of the war work done by women:

> They might pick some lint, and tear up some sheets,
> And make us some jellies, and send on their sweets,
> And knit some soft socks for Uncle Sam's shoes,
> And write us some letters, and tell us the news.

Walt Whitman described the specific use of lint in stanching blood flow in "The Wound Dresser" (1896):

> From the stump of the arm, the amputated hand,
> I undo the clotted lint, remove the slough, wash off the matter and blood,
> Back on his pillow the soldier bends with curv'd neck and side-falling head,
> His eyes are closed, his face is pale, he dares not look on the bloody stump,
> And has not yet look'd on it.

Flax was used to spin the strong thread used in manufacturing shoes, boots, tents, and other military supplies, but European fibers were difficult to obtain, and southern flax crops were insufficient to supply the Confederate army. Possible substitutes included the seed fibers of milkweed (*Asclepias syriaca*), also known as Virginia silk, which were used in the early nineteenth century for adding decorative stitchery to bonnets, collars, capes, and tippets.[23] Another option were the seed hairs of fireweed (*Epilobium angustifolium*), also known as rosebay willow herb, but neither was as strong or durable as flax. European nettles (*Urtica dioecia*) produce bast fibers similar to flax that were used historically for cordage and rope. In *Resources of the Southern Fields and Forests* (1863), Frances Peyre Porcher noted that nettle also was used to make thread and "white linen-like cloth of superior quality."[24] But its stinging hairs (histamine-containing trichomes) cause pain and inflammation on contact, which may have discouraged southern farmers from growing nettles or collecting them from naturalized populations.

Hemp (*Cannabis sativus*) is a weedy species that spread from Asia worldwide with human activity and agriculture. Settlers introduced the crop from England, where hemp fibers were used in manufacturing sails, rope, rigging, nets, and caulking for the seams of wooden ships. During the age of sailing ships, hemp was considered the finest fiber for canvas, a name derived from the name of the genus *Cannabis*. Like flax and nettle, hemp produces bast fibers which evolved as a stem adaptation for tensile strength; individual fiber cells can be about two inches long, and bundles of the flexible cells can be several feet in length. As an economic fiber, hemp fibers are four times stronger than cotton, and nineteenth-century products included carpets, cordage, paper, and heavy duty cloth. Some hemp was woven into coarse fabrics used to make workers' clothing and cloth bags for packing cotton.

Hemp was often favored over flax because fields yielded three or four times as much usable fiber per acre with less work. The calcium-rich soils of Kentucky and Missouri were particularly suited to antebellum hemp crops, but the federal Commissioner of Agriculture predicted that hemp would thrive in cooler climates and might do well in Midwestern states with minimal labor required: "Under favorable circumstances the crop makes its appearance in a few days, and with proper sun and moisture it rapidly covers the ground. From seed time until harvest, the laborer has only to watch its almost magic growth from day to day."[25] Retting resembled the process used in processing flax. Harvested stalks were spread on the ground where they were exposed to moisture and microbes, or alternately they were submerged in streams or ponds where the soft tissue surrounding the hemp fibers quickly decayed. Yet despite its weedy growth, hemp became known as a difficult crop to cultivate; fields were easily damaged by hail storms and wind, and small farmers lacked the manpower to harvest, ret, and bind large quantities of hemp fibers into sheaves.

Wartime ingenuity and shortages resulted in the use of several alternatives to flax and hemp. The long, fibrous leaves of yucca (*Yucca filamentosa*, commonly known as Adam's needle, bear-grass, or Spanish bayonet) were used to hang meats in smokehouses, which suggested the fibers as a possible alternative to linen and hemp. There were optimistic reports of wartime cordage and twine spun from yucca fibers.[26] By the Civil War, yucca was commonplace on farms and homesteads, and Cherokee healers also planted and used the rhizomes to prepare poultices for inflammation. The plants were native to beaches and sand dunes in the South, and they were easily propagated from root cuttings. Processing yucca required ingenuity using available machinery; as a wartime alternative to hemp, the leaves were boiled until softened and then pressed in a sugar cane mill, which released the useful fibers.[27]

New Zealand hemp (also known as New Zealand flax, *Phormium tenax*) was cultivated as both an ornamental garden plant and fiber source.[28] Its long leaves yield fibers that can be several feet in length, and these were used as a wartime alternative in making cordage and cloth. The plants tolerated poor soils and shade and were winter hardy in the Deep South. Other hemp substitutes included the stem vascular tissue of milkweed (*Asclepias syriaca*) and goldenrod (*Solidago canadensis*), common native species with weedy growth and widespread North American ranges.

Textiles, Clothing, and Uniforms

The Confederate states produced most of the global cotton supply, which provided textiles for clothing and other needs worldwide. Simple cotton dresses were a fashion that began with the muslin *robe de gaulle* worn by Marie Antoinette in a portrait painted in 1783 by Élisabeth Louise Vigée Le Brun. The gown attracted attention for its provincial style, and the popular style fueled demand for American cotton to supply the European market, coinciding by chance with the invention of improved cotton gins. Nineteenth-century fashions were elaborate in design, and inexpensive cotton cloth (as much as fifteen yards) was often favored for hooped skirts and dresses. Thrifty cotton fabrics included calico; fustian, a linen and cotton blend; and linsey-woolsey, woven with a cotton warp and woolen weft as a way to extend limited wool supplies. To give light weight cloth more substance, cotton clothing was often laundered with starch, which was one of many commercial products scarce in wartime. Southern women resorted to homemade versions of starch from unripe corn, sweet potatoes, Irish potatoes, or wheat bran.[29]

Civil War–era bonnets for women and girls were woven from a variety of botanical materials, including straw, corn shucks, palmetto leaves, and pine needles. In the case of grasses such as straw and corn, the stems and leaves contain silica deposits, an adaptation that strengthens tissues mechanically. Palmetto leaves produce tough vascular tissue, and pine needles develop layers of strengthening tissue (sclerenchyma), making them suited for bending, weaving, and exposure to weather. Inexpensive buttons and other ornaments were fashioned from available materials including wood, seeds, pits, and gutta percha, a thermoplastic latex harvested from *Palaquium percha* trees native to Malaya.

Shirts and undergarments were also made from cotton, including the corsets worn by women of all classes. These were fashioned from coutil, a tightly woven cotton cloth made specifically for the corset trade. Coutil resisted the stresses placed on cotton fibers by tight lacing and stays, the inserted supports made of whale bone or the stems of cane (also known as reed, *Arundinaria* spp.; see Chapter 3). While not true woody tissue, cane stems are impregnated with lignin, the hardening substance of true wood, and long strips of the stem tolerated tensile bending in corset designs without splintering.

While cotton was grown in the South, the spinning and weaving of fibers were the work of northern mills. In Confederate states, the blockade resulted in wartime cloth shortages, and any available fabric was used to make military uniforms. As a matter of routine, women repurposed clothing and household fabrics or used homespun cloth for wartime garments. Despite stringencies and extra work, southern women found time and resources to sew Confederate-themed flags, scarves, and clothing items from scraps and repurposed fabric, and many learned to spin and weave. Spinning wheels, spinning jennies, looms, needles, and thread became valuable household tools, and advertisements encouraged women to take up weaving as a patriotic wartime task. Many homesteads had flax wheels made from maple and other woods, and these were put back in action. Some devised their own machinery for spinning and weaving, as reported in the *Memphis Daily Appeal* (November 13, 1862), "A lady living five miles north of Ozark, Arkansas, with an axe, a saw, a chisel and an auger, made herself a loom out of oak rails, upon which she now weaves eight yards of coarse cotton cloth a day. The thread is furnished by Major N.B. Pearce, and woven into cloth for army purposes. Think of that, ye effeminates who loll on a sofa or carriage cushions and complain."[30]

A remnant of homespun textile production persisted after the Civil War when spinning wheels were repurposed into furniture. At the end of the nineteenth century, a period of Colonial revival emerged from the difficult years of the Reconstruction era, a time when Americans reflected on the history of the Founding Fathers and early colonial homes and traditions. As postwar industrialization spread, spinning wheels became superfluous, but homes needed furniture. Thus during the 1880s many flax wheels used during the Civil War were disassembled and the parts incorporated into Windsor-style chairs of pseudo-colonial design. Chair backs and arms were fashioned from the wooden wheels, and the turned spindles were repurposed as chair legs.

Clothing for enslaved people was made from cheap manufactured or homespun cotton cloth, with emphasis on durability rather than style. Workers who were still on plantations during the war did the spinning and weaving to produce cloth for household use and also for their own clothing, work that was done at the main house or during the evening in their cabins. Large plantations often employed seamstresses who made garments for both enslaved people and planters' families, sometimes with the aid of a sewing machine. Several of the WPA narratives include references to the work of enslaved women who made homespun cotton, osnaburg cloth from flax, and fabric for Confederate uniforms.[31]

Women spun a mixture of cotton fibers and cow hair from tanneries to make wartime necessities such as socks and blankets.[32] Blankets were woven from Spanish moss (*Tillandsia usneoides*), a bromeliad familiar as an epiphyte on trees in the Deep South (see Chapter 2), and these eventually became standard equipment for Confederate troops. In violation of military policy, clothing and bedding were among the items plundered or destroyed by Union troops, which added to shortages and hardships.[33] Southern women replaced these goods with homespun, homemade equivalents, and newspapers championed their efforts. As noted in *The Southern Confederacy* (November 7, 1862), "The hum of the spinning wheel and clank of the loom greet our ears and vision wherever we go. The instruments of machinery predominate, the piano retiring, and, for the present, being a useless instrument. Young and old ladies are exercising great diligence in spinning, weaving, and supplying clothing for the soldiers for the ensuing winter. They are entitled to the praise of the soldiers, and should receive, at home, every encouragement in their noble work."[34]

At the beginning of the Civil War, soldiers often wore what they had at hand, in many cases the uniforms of local militia groups. Military uniforms made use of the available fibers, and thus many Confederate uniforms varied widely depending on the supplies on hand, and many cotton growers held back cotton supplies for making uniforms for state militias rather than relinquishing supplies to the government. To conserve fibers and cloth, designs changed from long frock coats to short shell jackets. These coats and jackets were usually cotton-lined, but outer layers were often sewn from linsey-woolsey or woolen cloth. Another sturdy gray-blue cloth that combined wool and cotton was known as jeans or Virginia cloth, which prisoners in Louisiana used to make uniforms for Confederate troops.[35] Some versions of the cloth may have been entirely cotton and resembled denim. Much of the necessary sewing was also done by women, either as workers or as volunteers in aid societies. By September 1861, the Schuylkill Arsenal in Philadelphia employed 3,000 seamstresses who took pride in making uniforms sturdy enough for army use. The meager income earned by wives and mothers kept some soldiers' families from resorting to almshouses for survival during the war years.[36]

Once issued their uniforms, soldiers were responsible for necessary repairs. Many carried a "housewife," a rolled or folded cloth sewing kit outfitted with scissors, needles, thread, patches, and buttons. Families supplied many essential items, including home-sewn and knitted clothing; farm and homestead journals suggested sewing projects to benefit troops, especially Confederate soldiers who wanted for many necessities. Waterproof cloaks that doubled as blankets and outerwear were made from oiled cloth (prepared from whatever was available, probably linseed oil and cotton or possibly silk cloth), layered with cotton batting, and sent to soldiers for winter use.[37] In an era of desperation, cotton, linen, and wool were recycled from worn clothing and woven into homespun versions of linsey-woolsey made from any available fibers.

Military Needs

Civil War tent production in the North first depended on sailcloth, the cotton canvas used to make ship sails. Also known as duck, the cloth was woven on power looms in New England textile mills that specialized in cotton sailcloth, and it replaced the flax-based canvas used during the early 1800s. The dense weave protected against rain and retained warmth from fires and body heat.[38] Cotton duck was also favored in the South, but locally produced canvas probably also incorporated flax and hemp, depending on availability. Some captured tents were used, and some Confederate troops abandoned tents altogether in favor of shelters improvised from natural materials.

Early in the war both Union and Confederate armies encamped in Sibley tents, a bell tent design invented by Henry Hopkins Sibley for military use in the field. Each conical tent was twelve feet high and sheltered twelve men, although the original patent noted that they were intended for groups of twenty.[39] For ease in transport, each was supported by a collapsible iron pole and tripod, which doubled for use in cooking over a central fire pit. Ventilation occurred through an opening at the top, but there were widespread complaints about interior air quality. Sibley tents were used by the U.S. and British armies for several years, and the U.S. Quartermaster ordered nearly 44,000 for manufacture during the Civil War. However, the large size proved cumbersome, and after a year the Sibley tents were replaced by smaller tents that were comparatively lightweight and easier to transport. Federal officials later refused to pay Sibley or his heirs the royalties due for use of his patent, presumably because he served as an officer in the Confederate army.[40]

Sibley tents were used by Union and Confederate armies early in the war; they were made from canvas woven from cotton, flax, and hemp, depending on the fibers available. The bell tent design was supported by an iron pole and typically sheltered twelve men (*The Medical and Surgical History of the War of the Rebellion*, 1870–88, Vol. 3).

Wartime canvas was also used to make wedge tents, six feet of canvas draped over a horizontal pole and staked to the ground to make a shelter for six men. Shelter tents (often known as dog tents) were a smaller option and consisted of two canvas or twill (also known as drill) pieces that buttoned together to suspend over a ridgepole. They sheltered two men (each carried a tent half with their equipment), but they had to forage for branches or fence rails to serve as the ridgepole. Smaller tents for enlisted men avoided the need for wagons and horses to transport large numbers of canvas tents and poles.

Gum blankets were made of canvas covered

with rubber (see Chapter 8) on one side, and these were used as ground cloths or suspended for shelter. In the South, an improvised equivalent was described as "more pliant than India-rubber" and made by waterproofing cloth with beeswax dissolved in turpentine made from pine (see Chapter 8).[41] In contrast, officers utilized wall tents, with sufficient room for cots, field desks, and chairs; the same canvas structures were used for hospital tents.

Although the role of mosquitoes as a vector for malaria was not yet understood (see Chapter 5), biting insects were a continuous bane and annoyance. Standing water at encampments provided breeding areas for mosquitoes, which often moved with troops on the march. Confederate soldiers packed loosely woven cotton cloth to use as insect netting, a practical idea that was already in use on ships stationed in tropical regions. Some actual netting was imported from England, but the blockade interfered with shipments. Later in the war Confederate troops improvised with homemade cotton mesh, but the machine-made netting supplied to Union troops by mid-war proved more durable.

Ingenuity was essential with limited supplies and treacherous conditions. During winter encampments, some soldiers excavated the earth under their tents, constructed log walls, and layered canvas and oiled cloth as roofing (see Chapter 8). Other field modifications included installing an improvised floor under the tent, which was raised above ground level.[42] Due to ongoing canvas shortages in the South, Confederate soldiers often made do with makeshift shelters of oiled cloth rather than standard tents. In fact, some officers took pride in avoiding tents and the hardship of transporting heavy equipment, preferring instead to have their men improvise shelters from boughs and woody branches, as noted above. The imagery persisted in the canvas backdrops used in period photography, paintings on cloth that often featured soldiers' tents, shelters, and scenery.[43]

Haversacks were twelve-inch square bags with shoulder straps and an inner compartment or bag for bread. Most were made from heavy cotton cloth or canvas made from linen, hemp, cotton, or a mixture of fibers; they were used to carry food, eating utensils, and personal possessions. Haversacks were homemade or military issue, and many Union versions had an outer waterproofing layer of tar, the viscous material that can be isolated by destructive distillation from botanical sources such as peat, wood, and coal. Along with various types of tents, haversacks were among the military goods manufactured by prisoners such as those at the Louisiana State Penitentiary.[44]

Cotton flammability was dangerous in clothing, especially during an era of open fires and hearth cooking, but cotton was cheap and favored for many garments and insect netting. *Scientific American* reported on treatments developed by the French chemist Joseph Louis Gay-Lussac, in which plant fibers were rendered inflammable by saturating the cloth with a solution of ammonium chloride and ammonium phosphate. However, the frequency of its use and need for renewal is unknown.[45] The explosive known as gun cotton resulted from an accidental spill of nitric acid that was cleaned up with a cotton cloth, resulting in the inadvertent synthesis of highly flammable nitrocellulose. During the Civil War, gun cotton was packed into artillery shells, but its unstable nature caused reluctance to use it widely as an explosive. Perhaps ironically, gun cotton also had a medical use in making a substance known as collodion, an adhesive plaster used during the Civil War as a wound dressing. Collodion was compounded from potassium nitrate, sulfuric acid, gun cotton, and ether, and it resulted in a plaster "with wonderful tenacity, and forming a nearly transparent skin, airtight and nearly impervious to water."[46]

Medical uses of cotton increased during the course of the war, with clumps of raw fibers replacing cotton bandages when the latter were scarce. Although many doctors and nurses did not fully understand bacterial infections, cloth bandages were boiled to remove blood, which coincidentally sterilized them. Raw cotton fibers were routinely baked prior to use, which also prevented some infections.[47] Cotton was also used as a substitute for the surgical lint (linen fibers) used to dress wounds; the oils on cotton fibers rendered them less absorbent than cloth, but they may have hastened the process of heating and charring the fibers, which was the standard method of preparation.[48] Other medicinal uses of cotton included the use of the seeds as a quinine substitute and the roots for contraception (see Chapter 5).

Silk

Silk is produced by silkworms, the caterpillars of *Bombyx mori*, and is harvested from their cocoons. The strands consist primarily of fibroin, a protein similar to keratin, and the long fibers can be carefully unspooled when a cocoon is soaked in warm water. Although it is not a botanical fiber, silk production is fueled by plant tissue; silkworms feed on mulberry leaves (*Morus* spp.), and successful silk production depended on rearing both species—silkworms from moth eggs and mulberry trees from seed or cuttings.

The silk industry began in ancient China, but entrepreneurs carried the necessary skills for rearing silkworms to Europe and America. King James I shipped silkworm eggs and Asian white mulberry seeds (*Morus alba*) to Virginia with the expectation that colonists would produce silk for England, but colonial planters turned to tobacco crops after failed attempts at silk culture. There were also attempts to rear silkworms on native red mulberries (*Morus rubra*), but the Asian and American species soon hybridized, and some mulberries became locally invasive. During the Civil War, soldiers ate the berries and so helped to disperse the seeds further.

New England had thriving silk industries during the first half of the nineteenth century, including the Cheney Brothers of Manchester, Connecticut, where workers cultivated mulberries, raised larvae, processed cocoons, spun thread, and wove and dyed silk cloth. In 1832, Samuel Whitmarsh of Northampton, Massachusetts, planted 25 acres with mulberry trees and built a cocoonery to house two million caterpillars. Oriental mulberries (*Morus orientalis*) replaced the Asian and native species grown prior to the 1830s; these trees produced two crops of leaves each year, an advantage in feeding silk worms, but many mulberry orchards died during the harsh winter of 1844. During the antebellum years, some small silk industries survived in the North and produced silk for cloth, handkerchiefs, ribbons, and thread.

In "An Essay on the Culture and Manufacture of Silk" (1847), H.P. Byram of Kentucky noted that Asian mulberries thrived on well-drained, deeply cultivated soil suitable for corn cultivation, and he advised planting the trees closely in areas where land sold at premium prices. He recalled that the seeds he obtained from Canton "produced a variety of plants," suggesting possible hybridization, and that the trees with the best leaf crops produced lower fruit yields.[49] He supplied Asian mulberries to a Harmonist community in Economy, Pennsylvania, where efficient silk production was one of their sources of income. Textile industries depended on a ready supply of American silk, which was popular for both military and civilian clothing. In Philadelphia, William H. Horstmann and Sons invested in specialized looms and plaiting machines to produce various silk braids and ornamental trimmings to the U.S. military.

During the Civil War, there was increased demand for silk for flags, sashes, medals, and mourning clothes, as well as for making oiled silk, the waterproofed fabric used to cover bandages and make outer garments. Silk thread was used to suture surgical wounds and ligate blood vessels during amputations, but it was in short supply for the Confederate army. Silk was valued as strong and nonabsorbent, and horsehair was the only likely wartime alternative when suturing silk was unavailable. Some have suggested that the practice of boiling horsehair (a common practice to enhance its pliability for suturing) may have prevented some infections, although this supposition remains undocumented.

Ailanthus silkworms (*Samia cynthia*) offered a potentially less labor intensive means of silk production, a plan promoted by the new federal Department of Agriculture. *Ailanthus* silkworms thrive on the leaves of tree of heaven (*Ailanthus altissima*), the Chinese import originally prized for its foliage and growth (see Chapter 6), but eventually despised for its malodorous scent and invasive growth. The insects were introduced intentionally to the U.S. with the plan that the silkworms would complete their life cycles and produce cocoons outdoors without the need for indoor rearing. The *Report of the Commissioner of Agriculture for the Year 1862* offered detailed suggestions for this alternative means of silk production, suggesting that it could provide not only valuable fibers, but also "employment well adapted to the poor, or the aged, or the very young, who are not capable of performing any severe labor."[50] Although the caterpillars survived as far south as Georgia, the plan to culture *Ailanthus* silkworms went no further in Confederate states, despite the need for silk.

During the first two years of the Civil War, hot air balloons were in use by Union and Confederate armies to surveil opposing troops and direct artillery fire during battles. In particular, the Union army invested heavily in surveillance and had a balloon corps tasked with spying on Confederate moves. Union balloons used silk imported from India to make the outer envelope, while some colorful Confederate balloons were sewn from unused dress silk (not repurposed from dresses as some legends have repeated). Cotton was also used, both for ropes and in making the envelope of some Confederate balloons. In addition, hot air flight required bulky equipment to generate hydrogen, which was pumped into the balloon envelope sealed airtight with a varnish coating. Its ingredients are unknown but probably

Hot air balloons were tethered in place and used for military surveillance during the first two years of the Civil War. Union balloons were made from silk imported from India and varnished to seal the outer envelope, while some Confederate balloons were sewn from dress silk (*Harper's Weekly*, 1861).

consisted of pine resin dissolved in a solvent, which hardened into a yellow-tinted varnish impervious to hydrogen gas.

Military balloons reached heights of 1,000 feet and provided the crew (up to five men) with views of several miles. For operation aloft, they relied on civilian aerialists, who were often itinerant performers who claimed expertise in flight or chemistry. Controlled by the use of sandbags as ballast and hydrogen gas for lift, balloons in free flight were tricky to control and potentially hazardous. Thus the balloons were usually tethered to ground sites and naval ships, where they were used in controlled ascents, or occasionally navigated over battlegrounds. They tended to attract enemy fire because of their size, stationary location, and exposure. For instance, during the 1862 Peninsula campaign, Confederate troops fired routinely at the Union balloon as it ascended and descended over Yorktown, but failed to bring it down.

Paper

During the mid-nineteenth century, high quality paper was made from repurposed linen and cotton rags mechanically reduced to pulp, and chronic shortages even led to the use of cloth recycled from Egyptian mummies.[51] At the beginning of the Civil War, northern states had over five hundred paper mills, compared to about two dozen in southern states, and most printing presses and supplies such as type and commercial ink were also more readily available in the North. Paper demand increased in wartime, and at a time when southern periodicals sought to encourage Confederate resilience, paper shortages resulted in shorter and fewer publications. Some newspapers collected rags for papermaking, but cotton cloth was in high demand for bandages, and cotton bales were routinely burned to prevent capture by Union troops. As conditions became more desperate in Confederate states, cloth shortages made the manufacture of new paper almost impossible, and used paper was repurposed to make insulating layers in clothing and bedding.

As a result of ongoing paper shortages, newspapers and journals merged, published bimonthly or sporadically, or suspended publication entirely. These included farm journals with occasional wartime issues and daily newspapers such as the *Charleston Courier*, which by 1865 was reduced to a single four-column sheet with limited wartime messages and battlefield reports. Some Confederate newspapers resorted to printing on unused wallpaper, ledger paper, colored paper, and wrapping paper.[52] The May 20, 1862, edition of the *Natchez Courier* was printed on deep green paper, which editors explained to their readers: "The publication of a daily newspaper, just at this time, is rather slim business. With our exchanges from New Orleans and from Vicksburg, entirely blockaded out by the enemy, it is sometimes extremely difficult to obtain paper from those points.... We trust our readers will bear patiently with us until a change of affairs shall occur."[53] The *Charleston Mercury* (January 18, 1862) summarized the situation well: "The newspapers on all sides begin to show the marks of the scarcity of paper. The New Orleans Picayune has discontinued its evening edition; the Delta continues to publish twice a day as formerly, but uses only a half sheet; the Savannah News comes to us printed on brown paper; and among the journals generally half sheets and all the colors of the rainbow, are rapidly growing epidemic."[54]

Books were also in short supply in Confederate states, and bibles were among the household goods plundered by Union troops that were difficult to replace. Soldiers on both sides carried bibles and hymnals into battle, including editions such as the *Soldier's Pocket Book* and *The Soldier's Hymnal*, which offered comfort in the form of psalms, prayers, scripture lessons, and hymn verses. The U.S. Christian Commission, a volunteer Protestant charity, was organized to provide bibles, hymnals, writing paper, and postage stamps to Union soldiers, while Confederate troops relied on whatever southern publishers could produce. Prior to the Civil War, most books were imported from England or shipped from northern states, but publishers in Atlanta, Mobile, Richmond, and other cities managed to print wartime editions on brown paper and bind them with wallpaper or other materials on hand. Titles included military histories, song books, textbooks, and novels, but stock was scarce, as were supplies of account books and stationery. Civilian and military diarists resorted to using wrapping paper and overwriting previous entries, and prisoners of war kept diaries on scraps of paper using ink improvised from rust or other dark substances.

Papermaking had a history of innovation in the event of fiber shortages, and there was wartime interest in alternate sources of botanical fibers for making paper. During the 1820s, John Shyrock had experimented with macerating the stems and leaves of wheat, barley, oats, corn, and buckwheat at his Pennsylvania paper mill, concluding that rye and wheat were the best alternatives. Combined with rag fibers, they yielded a strong paper which was suitable for sacks, wrapping, and printing newspapers. However, the use of dried stems and leaves resulted in impurities that caused inconsistent color and texture, a problem solved by the invention of machinery that removed large pieces of plant tissue and crushed the remaining fibers between heavy rollers. The process was known as the stable fiber method, and rye-based paper produced using this way was cheap enough to print New York daily newspapers during the Civil War.[55]

Manila paper was developed during the cash-strapped years of the Panic of 1837. James Whitney and Lyman Hollingsworth used discarded remnants of rope and canvas from fishing boats as a source of fibers for their Connecticut paper mill. The resulting mixture of fibers used for canvas and cordage included jute (*Corchorus capsularis* and *C. olitorius*), a coarse fiber imported from India; manila fibers derived from the mature leaves of a banana relative (*Musa textilis*) from the Philippines; as well as flax and hemp. The fiber mixture produced the smooth, cream-colored product which became known as manila paper, and it suggested paper-making strategies for wartime use. During the Civil War, the Department of Agriculture described papermaking experiments in Europe using corn shucks, well known for their fibrous anatomy, particularly at the Schlögelmuhl in Austria where numerous types, weights, colors, and textures of corn-based paper were produced with high success.[56] Earlier patents had been issued to papermakers for producing stock from both husks and cobs, and high wartime prices for rag paper pointed to reviving the idea of paper made from an abundant native fiber, but the advice was ignored by large scale paper manufacturers.

In *Resources of the Southern Fields and Forests*, Francis Peyre Porcher recommended that Southerners experiment with familiar fibrous plants as alternative paper fibers, including the leaves of agave (*Agave americana*) and stems of bulrushes (*Typha latifolia*), sunflowers (*Helianthus annuus*), and sorghum (see Chapter 4). He also mentioned the bark of the native red mulberry (*Morus rubra*) and the paper mulberry (*Broussonetia papyrifera*), introduced from Asia as an ornamental tree but which also had traditional uses in paper-making.[57] Whether many had the time or resources to experiment on a large scale with alternative fibers is doubtful, but paper shortages in both the North and South affected postwar paper manufacture. As early as the 1820s, some papermakers had experimented with using wood pulp and sawdust; by the end of the war American paper mills began to follow the German practice of using wood pulp in paper manufacturing.

Dyestuffs and Pigments

Many botanical dyes have an ancient history in the European flora, and more dyestuffs yielding pigments were discovered growing in North America. Dye plants were among the household species cultivated in early American herb gardens, and by the time of the Civil War, knowledge of dyestuffs included several plants used by Native Americans. Despite the bright pigmentation of flowers and fruit, permanent plant dyes usually originate in roots or leaves; red, blue, and purple anthocyanin pigments do not dye fibers permanently. Even reliable dyes differ in the extent to which they are fugitive, meaning that they fade with exposure to sunlight, soap, and water.

By the mid-nineteenth century, many botanical dyestuffs were purchased rather than collected or cultivated. Synthetic dyes were also known by the 1850s, beginning with William Henry Perkin's serendipitous discovery of mauve, based on an aniline residue remaining from attempts to synthesize quinine to treat malaria. However, as a result of Civil War era shortages, southern women in particular often prepared their own dyes from available plant material. A knowledge of practical chemistry was essential in dyeing textiles, methods often acquired through trial and error. Key techniques included the use of mordants, which were usually metal ions that chemically linked dyes with fibers. The use of iron, tin, and copper pots often sufficed, but folk wisdom often included the addition of rusty nails to dye solutions. Prepared with a mordant, dyes were often more colorfast, and the range of possible hues expanded depending on the metals used. The addition of acids and bases also affected the colors imparted by slightly acidic plant dyes. Acids such as vinegar supplied $H+$ ions that were attracted to fibers, which in turn attracted more dye molecules and deepened colors. The addition of a base such as ammonia (often in the form of urine) to the dye bath resulted in fewer dye molecules on the fibers and resulted in paler tones.

Indigo was well known as a plantation crop, but Abolitionists avoided cloth dyed with true indigo (*Indigofera tinctoria*, see chapters 1 and 3) because enslaved labor was involved in its cultivation and processing. Traditions nevertheless evolved from deep blue indigo tones. Enslaved people used the dye to tint paint "haint blue," a pale tone that was thought to protect homes from evil spirits. Haint blue was used in outside decoration, and its hue was part of the indigo-inspired belief system which included bottle trees, long believed to capture evil spirits (see Chapter 5). Indigo preparation used enslaved labor in the lengthy processing steps of leaf fermentation in water, mechanical agitation, and precipitation of the dyestuff (indigotin) using limewater or lye, followed by draining and drying steps.[58] Southern women scaled down the process to make blue dye for homespun using wild indigo (*I. caroliniana*), a native species that they collected locally. Porcher described the coloring yielded by wild indigo as "equal to the commercial indigo, which dyes a beautiful and lasting blue."[59] Dye baths were prepared by pulverizing indigo cakes in water, adding urine or wood ashes, and allowing bacteria to ferment indigotin to its soluble form for use as a household dye. Some may have experimented with other legumes including dyer's baptisia (*Baptisia tinctoria*) and bastard indigo (*Amorpha fruticosa*) to precipitate similar dyes. Other blue dyes were prepared from the shoots of purslane (*Portulaca oleracea*), a weedy, edible garden plant introduced from southern Europe; leaves of black locust (*Robinia pseudoacacia*); and the bark of white ash (*Fraxinus americana*).

The berries of pokeweed (*Phytolacca americana*) yielded a crimson dye that also served as an ink. The same species was used as a potherb, but it is now known that all parts of the plant contain phytolaccin, a mitogen that increases cell divisions and may cause blood abnormalities merely when plant parts are handled; possible ill effects may not have been attributed to handling or consuming the plants.[60] The leaves of Saint John's wort (*Hypericum perforatum*) yielded a red dye, and its flowers yielded a yellow dye. Originally introduced as a European household herb, Saint John's wort was used to treat skin sores on horses and cows, female problems, and hysteria, and it was also "thought to cure demoniacs."[61]

When iron was added to a dye pot as a mordant, the inner bark and nut shells of butternut trees (*Juglans cinerea*) produced the gray dye that was the standard color of many militias in the South. The combination was used to dye cloth for the shell jackets worn by Confederate soldiers, sometimes called "butternuts" by Union troops. The same species yielded tan and brown, depending on the mordant and whether the nut husks were added to the dye pot. The leaves of bald cypress (*Taxodium distichum*) produced a deep cinnamon-colored dye when boiled in water for several hours. Native to southern swamps, the trees were felled for their timber (see Chapter 8), and the leaf dye was a serendipitous discovery. Experienced dyers understood considerable practical chemistry and willingly experimented; when it was available, tropical American logwood (*Haemotoxylon campechianum*, a legume native to Mexico) was used to deepen the tones of other botanical dyes; combined with quercetin from the bark of black oak (*Quercus velutina*), it produced a deep black dye. Black oak alone produced bright yellow when alum (aluminum sulfate) was used as a mordant.[62]

At a time when black dyes were in high demand for mourning clothes, Porcher described several alternatives, including the bark of sweet gum (*Liquidambar styraciflua*) combined with copperas (ferrous sulphate) which "yields a color nearly black," and the bark and roots of sumac (*Rhus glabra*) which "furnished the country people here … a most substantial dyestuff (a brilliant black)."[63] According to a correspondent to the *Southern Cultivator*, various iron-containing decoctions of sassafras root (*Sassafras albidum*), elderberries (*Sambucus nigra*), and the bark of willow (*Salix* sp.) and maple (*Acer* sp.) also yielded permanent black dyes.[64] Some may have been an improvement over the black walnut husks (*Juglans nigra*) that many southern women used to dye their mourning clothes, at a time when ready-made black cloth was nearly impossible to purchase.

8

Timber and Wood

The Timbered Landscape

Although by the 1860s over 150 million acres had been cleared for agriculture, the remaining forests in the North and South offered rich natural resources. Land was a cheap commodity, and farms and plantations intentionally included forested tracts for eventual clearing. Southern agriculture in particular involved cycles of forest destruction and reforestation. As cultivated fields lost soil nutrients, they were abandoned; new land was cleared, and old fields regrew with shrubs and young pines, a process now known as secondary succession.

The southern climate supported diverse tree species and promoted rapid woody growth, and local economies included industries based on timber and wood products. Before the Civil War, Confederate states supplied timber nationwide, including valuable hardwoods and the raw materials for naval stores. The latter originally included all of the botanical products used in shipbuilding, including fibers, timber, tar, and pitch, although by 1800 the concept of naval stores referred only to the products derived from pine resin and wood—tar, pitch, rosin, and turpentine.[1] Demand for all of these goods was great. Nationwide growth during the nineteenth century required timber for buildings, barrels, wagons, railroad tracks, fuel, and charcoal. Shipyards required planks, masts, and pine products, and the tanning industry depended on bark to preserve leather. Yet despite the high demand for timber, the regions most affected by the Civil War were still largely forested in the 1860s and included tracts of virgin forests.

Hardwood forest trees in the eastern U.S. included several species of oak (*Quercus* spp.), beech (*Fagus grandifolia*), and chestnuts (*Castanea dentata*), which had not yet been affected by the chestnut blight. Conifers are softwoods, and these included longleaf pine (*Pinus palustris*) which grew in coastal regions of the Carolinas and Deep South. Settlers understood the hardwood and softwood properties needed for various uses, including practical knowledge of wood density and strength and the tendency of timber to shear or compress. On a microscopic level, conifer wood consists of tracheids, water conducting cells that are dead and hollow at maturity. Softwoods lack the flexible, thick-walled fibers that strengthen hardwoods, explaining why pine and other conifer timber is typically less dense and more prone to compression and damage. Wood strength is reflected by specific density, which can be measured by comparing the mass of equal volumes of wood and water. The specific gravity of temperate woods ranges typically from 0.35 (various pines, *Pinus* spp.) to 0.65 (shagbark hickory, *Carya ovata*). Lignum vitae (*Guaiacum officinale* and *G. sanctum*) is among the few woods with a specific gravity (1.32) higher than water (1.0), and it sinks when submerged. There is nothing comparable to lignum vitae among native timber, and Civil War supplies for mallets, pulleys, and shipbuilding came from European importers who acquired the wood from Caribbean and South American plantations.

As the dominant trees in many southern forests, chestnut and oak were used for split rail fences around fields and farms, but fencing provided Union troops with a wartime source of timber for fires and shelters. As a result, fields were left unprotected and animals roamed freely, jeopardizing the food supply for both civilians and soldiers. With a natural range that extended from southern Maine into Appalachia, chestnut trees grew to maximum size in high rainfall areas such as southern hollows at elevations over 3,000 feet. In many areas, chestnuts comprised 20 percent or more of all forest trees, and they grew to robust dimensions which exceeded seven feet in diameter and 120 feet in height. The nuts were a communal resource that provided forage for pigs and a wild crop for families to barter or sell during lean times. With a specific gravity of 0.48, chestnut wood was light weight, strong, decay-resistant, and easily split, which accounts for its diverse uses in fences, log cabins, furniture, and coffins.

American chestnut trees provided timber for fences, cabins, and coffins, and the nuts were bartered or sold by cash-strapped families. During the mid-nineteenth century, they accounted for one-fifth or more of the trees in many areas from southern Maine into the Appalachian Mountains (author's collection).

Oaks had a similar wide range, and they grew in every type of hardwood forest affected directly by the Civil War. Acorns were an emergency food for both civilians and soldiers (see Chapter 4), and mammals and birds (including wildlife hunted as food) depended on these wild nuts as a staple in their diets. Settlers soon learned that some acorns were preferable; those from white oaks mature in a single season and contain fewer tannins (and so are less bitter) than red oak acorns, which require two years to mature. Experience also revealed the various uses of oak timber. Red oaks (specific gravity = 0.56) and related species have porous wood that was unsuited for barrels and shingles; white oaks (specific gravity = 0.60) produce dense wood that was valued for construction, cooperage, wagons, and tool handles.

At the beginning of the Civil War, wood was such a valued commodity that there was concern about widespread deforestation and its possible effects on the economy, landscape, and climate. In 1862, the *Report of the Commissioner of Agriculture* warned, "Our natural forests are rapidly being destroyed, and it is admitted that the destruction of forests tends to lessen the moisture both of the atmosphere and the soil" and noted the effects of clearcutting in other countries.[2] The advice to growers and farmers to plant "belts and groups, in masses, of hardy, suitable trees, in the vicinity of orchards and gardens" was forward-thinking, but unlikely to happen on a large scale when many women were managing alone on family farms.[3]

Nor could the effects of war be overlooked. Timber was cut for military construction and fuel, trees were scarred by gunfire, and forests burned. An article in the 1865 *Report of the Commissioner of Agriculture* warned of a desolate landscape resulting from the combined effects of timber damage and demand. Recommendations encouraged a nationwide program of tree propagation and suggested widespread replanting with native and non-native trees, including the cold-tolerant European larch (*Larix decidua*), which naturalized widely in the North.[4] Postwar timber management occurred when the Reconstruction-era Department of Agriculture appointed a special agent tasked with studying the extent and condition of U.S. forests, a position that evolved into the Bureau of Forestry and U.S. Forestry Service.

Farms and Homesteads

The demand for household wood in the nineteenth century cannot be overstated. From furniture and fuel to barrels, baskets, baking powder, and soap—all required access to timber. Vernacular furniture was crafted from wood on hand, depending on the preferences of the craftsman, and items crafted during the war were typically simple and often crudely finished. The dense wood from large ash trees (white ash, *Fraxinus americana*, and green ash, *F. pennsylvanica*) was used for sturdy bedsteads, tables, and chairs; seats were woven from ash or oak splints (made by pounding logs which separated at the annual rings into woody strips) or the native bamboo known as cane (*Arundinaria* spp; see Chapter 3).[5] Bedsteads were in particular demand as homes were destroyed and households relocated, and mattresses were filled with whatever materials were available, including straw, dried leaves, corn husks, cotton, and Spanish moss (see Chapter 2).[6] Beech leaves (*Fagus americana*) were considered particularly desirable, with a scent described as "grateful and wholesome," and they were available for replenishing mattresses each fall.[7] In warm areas, palmetto leaves were used in mattresses but were first prepared by shredding, boiling, and sun-drying to eliminate possible vermin.[8]

Tulip poplar (*Liriodendon tulipifera*; see chapters 2 and 6) was also widely used to craft furniture for southern homesteads. Sometimes known as yellow poplar, these are the tallest flowering trees in North America. In southern forests they grew to heights of 120 feet and diameters of five feet or more, and they were an abundant source of timber. The wood is strong and easily worked but remarkably lightweight, the result of the thin-walled fibers which strengthen the conducting tissue but result in low density. Although usually light-colored, the heartwood of mature trees was sometimes olive green in tone, the result of accumulated wastes, resins, and minerals. However, neither ash nor tulip poplar had the prominent grain or deeply pigmented heartwood desired by the planter class for fine furniture, but they were suitable for vernacular and regional pieces in which strength was valued.

Bald cypress (*Taxodium distichum*) are conifers indigenous to coastal swamps from Florida to Delaware, and in terms of timber mass, they were the largest trees in the eastern U.S. They are known for the woody "knees" that grow from submerged roots, described by Frederick Law Olmsted as "curious bulbous-like stumps" which were still visible in cypress swamps cleared for rice cultivation.[9] Once thought to uptake oxygen, the function of these emergent structures remains unknown. The deeply pigmented heartwood was used in both regional and high style furniture; color was considered in design, and craftsmen often used various woods such as cypress, walnut, and mahogany in a single piece of antebellum furniture. Inside plantation houses, cypress panels were one of the materials used to make ceiling fans known as punkahs which enslaved people lifted and lowered by cords to create a slight breeze in hot, humid air.

The interiors of many antebellum plantation houses were decorated in Rococo revival style, a complement to the European-style parterres in plantation gardens (see Chapter 6). Rococo revival began in England, gained popularity in France, and was soon adopted by U.S. furniture makers who catered to customers with social aspiration and money. Furniture in the Rococo revival style involved curvilinear shapes with intricate carvings and piercings, decorative elements that required expert craftsmanship to prevent wood from breaking or splintering. Several American furniture makers produced Rococo revival furniture, but John Henry Belter was particularly well known for his techniques and detailing. He was a German immigrant who worked as a cabinetmaker in New York from about 1840 until his death from tuberculosis in 1863, and wealthy buyers sought his elaborate furnishings. Belter's parlor furniture was particularly desirable, and before the Civil War large quantities were shipped to Southern states.

Various rosewood species (*Dalbergia* spp., native to Brazil and India) were valued for their varied coloration that included hues of deep reddish brown and black. The timber is dense and strong (some rosewoods have a specific gravity as high as 0.88), and it tolerated bending without splintering. Belter developed and patented techniques for working with rosewood using lamination and steam to achieve the desired forms. He probably began with the established practice of making three-ply panels for crafting furniture, a technique developed by eighteenth-century English cabinetmakers. However, Belter experimented further with plywood panels by gluing thin wood layers oriented at right angles to the grain; his patent "Improvement in the Method of Manufacturing Furniture" specified seven layers using exterior vertically-aligned rosewood veneers and five interior sheets of hickory, black walnut, or oak.[10] The veneers on doors and drawers were often matched at the center seam to make a symmetrical pattern of rosewood coloration.

Parlor sets and other items made from Belter's engineered panels were comparatively lighter and stronger than furniture crafted from solid hardwood. The panels were bent and molded using steam pressure, followed by carving designs in high relief. Other techniques involved gluing staves into large cylinders that were then molded with steam heat into desired shapes; one cylinder yielded the curved backs for several chairs. If necessary, wood blocks were glued to the surface to accommodate deeply incised patterns. Another Belter patent described a saw designed to cut the piercings for various carved patterns, particularly seatbacks.[11] His various carvings were often botanically inspired and included roses, morning glories, grapes, pomegranates, oak leaves, and acorns. These designs were most likely gleaned from British pattern books, but Belter adapted them to the Rococo revival décor of his American clientele, including many plantation owners for whom high style reflected prosperity.[12]

Timber provided for more basic daily needs, including fuel and illumination. Faced with tight coal supplies due to labor shortages, households in southern cities returned to burning wood in stoves and fireplaces. In some cities such as Mobile and Vicksburg, drastic conditions forced city officials to commandeer and distribute fair loads of fuel wood to residents.[13] Hardwoods provided the most heat, and trees harvested from a southern exposure were thought to provide better fuel, but wartime conditions allowed for little choice.[14] Wood shortages in cities drove prices high, even for green pine that burned poorly, and unscrupulous dealers learned how to stack logs to make cordwood appear more plentiful than it was. Some turned to charcoal when supplies were available. Its preparation involved the slow burning of virtually any softwood or hardwood under low oxygen conditions, and the resulting charred remains were used as fuel. Some may have experimented with various techniques for preparing charcoal on a small scale, including the English method of covering a wood pile with turf and allowing only minimal air to penetrate the pile during combustion. Charcoal was also known as an antiseptic, fertilizer, laxative, and porous material for filtering impure water, so a supply on hand was often useful.[15]

Desperate families resorted to stealing wood from railroad yards or making balls of sawdust and coal (the fossilized remains of Carboniferous forests) bound with clay. The dried balls could be used to sustain a fire once it started to burn.[16] Some civilians turned to peat-like alternatives such as the compacted dried leaves and rhizomes of bracken fern (*Pteridium aquilinum*, indigenous to every continent except Antarctica), a familiar practice in England since ancient times.[17]

Fireplace ashes were used to fertilize food crops, and they were also set aside for making soap and baking powder, both in short supply during the war years. Southern women became practical chemists in learning to extract lye and potash; the technique involved pouring water through sifted ashes, which yielded a lye solution. This liquid was boiled down, and the concentrated salts were melted to make pearl ash, which was used as an early baking powder (see Chapter 4).[18] When soap became scarce, housewives used ashes from straw, corncobs, and hickory and other woods as a source of potash, the various potassium salts that could substitute for lye in soap-making. In the South, enslaved people did the work of collecting ashes and making household soap, potash, and lye, and they used lye to prepare hominy from corn grains (see Chapter 4). The process was recalled by a former worker: "They saved the ashes and put them in a barrel and poured water over them and saved the drip lye and … made big pots of soap and cooked pots full of lye hominy."[19] To meet wartime demand, soap-making factories opened overnight in Confederate states, but shortages persisted although newspapers published recipes for making homemade soap from potash and grease. Animal fats were traditionally used in soap-making, but wartime meat shortages encouraged large scale experimentation with oils from cottonseed, castor beans, Chinaberry seeds (see chapters 2 and 5), and pine resin as possible fat substitutes in soap production.[20]

Most kerosene and candles were shipped from the North, and wartime supplies dwindled in Confederate states, thus civilians sought alternatives for illumination. Following tradition, some households burned pine knots, the remains of side branches embedded in mature wood which contain concentrated flammable resin. Known in colonial history as candlewood, the knots of pitch pine (*Pinus rigida*) produced a particularly bright light when burned. Other alternatives to kerosene lamps and candles included strips of cloth or paper saturated with pine resin, beeswax, or fat; a lamp oil substitute made from melted lard and turpentine; and the woody fruit of sycamore (*Platanus occidentalis*) or sweet gum (*Liquidambar styraciflua*) saturated with oil or lard. Farm journals suggested these options as well as "Confederate candles" made from a wick immersed in a mixture of beeswax and pine resin.[21] All were messy options, and the irony cannot be ignored of elaborate plantation house décor tainted by soot from burning candlewood, resins, and fats.

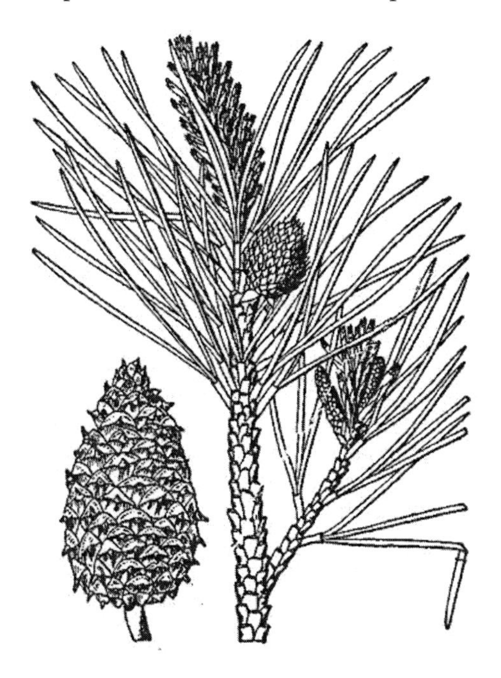

Southerners burned the resiniferous knots of pitch pine for household illumination when supplies of kerosene and candles dwindled. Other options included strips of cloth saturated with pine resin, beeswax, or fat and "Confederate candles" made from wicks immersed in a mixture of resin and wax (author's collection).

There was wartime interest in growing tallow trees (*Triadica sebifera*), a Chinese species already cultivated in southern gardens.[22] Its seeds have an outer layer (aril) containing stillingia oil, used traditionally in China for making candles and soap. A mature tallow tree produces annually 200,000 seeds that retain their viability for years and are easily dispersed by birds and water currents. Some Southerners may have pressed the oil and discarded the seeds, explaining the rapid spread of tallow trees as an aggressive invasive species in Florida and areas of the southern coastal plain. Compounds from tallow trees slow the growth of competing plants, an adaptation for survival known as allelopathy, which contributes to the rapid growth of tallow trees in warm climates.

Confederate states set up factories to make other household items previously shipped from the North, including wooden pencils and matches, which were advertised as being as good as pre-war supplies. Ink was also difficult to obtain, and recipes were passed around for homemade mixtures made from the tannins in oak, chestnut, and hemlock bark. When commercial printing ink was unavailable, wild grapes were crushed and used for printing currency.[23] The deep red and blue anthocyanin pigments are chemically unstable outside of acidic conditions, which may explain the extreme fading of some Confederate paper money.

Basketry was a traditional craft that often utilized strips of white oak wood to weave into baskets with various sizes and diverse uses. Oak splits easily along the large vascular rays (files of cells that move water horizontally in tree trunks), but wartime timber demand depleted oak populations in many areas of the South. Although wood strips from ash (*Fraxinus* spp.) and elm (*Ulmus* spp.) were also used for weaving, willow branches (*Salix* spp.) were often preferred for making sturdy baskets. The trees often colonized stream banks, and the branches were easily collected. European osier willows (*S. viminalis*) were planted to form dense hedges, and the long flexible shoots were preferred for basketry. According to the *Southern Cultivator*, "White oak is getting scarce in many localities, and we ought to raise plenty of the Osier Willow for Cotton baskets, willow cradles, clothes baskets, and the thousand other forms for which ornamental wicker-work is used."[24] Introduced osier willows naturalized widely in eastern states, sometimes hybridizing with native willows, and populations took root in damp soils.

In Confederate states, tools were in short supply during the Civil War, and various implements were routinely repaired because they were impossible to replace. The handles of shovels, hammers, axes, and other farm and household tools required wood that was dense, strong, and shock-resistant; common trees used for crafting replacement handles included beech, chestnut, elm, and oak. The dense wood of dogwood (*Cornus florida*) was used for tool handles and also for carving the spindle-shaped shuttles used with weaving looms (see Chapter 7).

Barrels were used in both household storage and shipping, and Civil War cooperage depended on any available wood that could be trimmed into staves. Supplies were plentiful enough to be used in devising punishments for imprisoned soldiers, who were sometimes forced to stand on a barrel for hours or parade while wearing a barrel with the ends removed. However, quality varied, and wartime barrels often could not contain liquids such as molasses, the result of poor craftsmanship or unsuitable timber. White oaks (*Quercus alba*) and their relatives were less porous than red oaks (*Quercus rubra* and related species), the result of microscopic tyloses that block cells in the xylem (water-conducting tissue). Tyloses form when nearby cells protrude into conducting cells and prevent water flow, and thus white oak barrels were generally regarded as watertight. Some woods such as American beech (*Fagus grandifolia*) tended to shrink as they dried, so instead were used in making barrels for dry storage. Other trees used for cooperage included chestnut (*Castanea dentata*), juniper (often known commonly as cedar, *Juniperus virginiana*), red mulberry (*Morus rubra*), and sassafras (*Sassafras albidum*); all produced strong wood that resisted decay, and the terpenes in juniper essential oils prevented insect infestations. Juniper was also valued for making rot-resistant canteens.

Military Needs

The Civil War predated the invention of Bakelite and other synthetic wood alternatives, and thus during the 1860s military equipment often required timber or forest products for its manufacture. Black walnut (*Juglans nigra*) was the most desirable wood for gunstocks because it is shock resistant, and its density (specific gravity = 0.55) and resistance to rot were also valuable attributes. High demand resulted in chronic wartime shortages of seasoned black walnut timber, which was also in demand for making knife handles, sword scabbards, coffins, and ships. Gunstocks crafted of the green wood tended to splinter, but thorough air-drying required years. Gun-makers sometimes substituted wood from butternut (*J. cinerea*), the walnut relative which provided the gray dye used for Confederate uniforms (see Chapter 7). The notoriously hard, dense wood of sugar maples (*Acer saccharum*, specific gravity = 0.63) was another alternative for gunstocks, but maple wood was better known for crafting military drums and fifes.

The heavy ramrods used to load cannons were crafted from sections of hickory logs (*Carya ovata, C. glabra*, and related spp.). The propellant was gunpowder (also known as black powder) in which the explosion was fueled by charcoal from incompletely combusted wood cell walls. There was high demand for charcoal, particularly the product made from cottonwood trees (*Populus deltoides* and *P. heterophylla*) which colonized southern riverbanks. Charcoal for use in gunpowder was also made from willow, dogwood, alder, hazels, and even the woody stems of hemp (see Chapter 7).[25]

Troops with canvas tents used them, but many encamped troops used any available logs to construct huts or lean-tos and crude beds layered with leaves, straw, or pine needles. Juniper groves were felled for their aromatic branches and logs for shelter construction (see Chapter 2). Bivouacked troops built winter cabins made by stacking and chinking logs with mud; even poor quality timber such as cottonwood and the trunks of palmetto palms (*Sabal palmetto*, not true wood because palms lack a vascular cambium) were used for these temporary structures. Floors were made from barrel staves, and roofs were improvised from thatched branches, often reinforced with canvas or oiled cloth (see Chapter 7). Boards from shipping crates were repurposed to build crude furniture. Large shelter tents were erected on log or lumber platforms, sometimes excavated into the soil and reinforced with wood salvaged from crates (see Chapter 7).[26] The demand for wood altered landscapes as trees were felled and stripped for military construction and fuel, and localized clear-cutting converted encampments into seasonal quagmires or dust bowls. Pathways through mud were lined with logs, and fuel wood became scarce wherever troops encamped.

Troops often used available logs to build shelters for encampment. As trees were felled for construction and fuel, clear-cutting converted military encampments into seasonal quagmires or dust bowls (*Harper's Weekly*, 1864).

Fences vanished (see Chapter 2), and military woodcutting details had to search further distances to find forests or woodlots to plunder.

Faced with leather shortages in the South, factories appeared that made wooden shoes for Confederate civilians and soldiers, although the soles were too inflexible to be worn while marching. During the war years, wooden-soled shoes became commonplace in southern states, but the practice was nothing new; enslaved people had worn clog-like shoes for years. The favored woods were tulip poplar and black gum (*Nyssa sylvatica*) from which soles were carved, sanded, and then "lined, and painted, and topped with leather."[27] Black gum wood was known to resist splitting, the result of a variation known as cross-graining (spiral arrangement of wood cells), which explains anatomically why the trees were also favored for making prosthetic legs. Leather uppers were attached to the soles with pegs cut from maple wood. Persimmon wood (*Diospyros virginiana*) had diverse wartime uses in clothing soldiers; leather boots were assembled on lasts made of its dense wood (specific gravity = 0.83), which was also used to carve shuttles for weaving cloth for uniforms. Some Southerners used flat persimmon seeds to improvise buttons, and the fruits were used as both a food and possible anti-malarial medicine (see chapters 4 and 5).

Camp chests and field desks with multiple compartments were made from oak, chestnut, and other hardwoods, often with single wide boards used for the top and sides. Commercial models were widely advertised to soldiers and their families; some were nothing more than a locked wooden crate for personal belongings, but Parr's Patent American Camp Chest for Army Officers was equipped with cookware, dinnerware, and camp stools, and it opened into a table for four men. However, officers may have valued field desks primarily for organizing papers, records, and letters while on the march. Some desks were crafted to suit an officer's personal preferences, and others were mass produced as freestanding and tabletop models with various arrangements of cubbyholes, slots, shelves, and writing surfaces.

Military prisons were often constructed in haste from available lumber, but unseasoned wood soon split, causing cold, damp conditions within the barracks. Combined with the hardships of minimal food, stagnant water, poor sanitation, and over-crowding, substandard construction contributed to high death rates at Civil War prisons, including the one built by the Union Army at Elmira, New York. The walls surrounding prisons were typically stockades constructed from vertical logs. The notorious Confederate Andersonville prison in Georgia had a simple pole fence known as the "dead line" erected several feet within the stockade. It was designed to prevent prisoners from scaling or tunneling under the stockade, and guards shot any prisoners who touched it. Confederate prisoners of war often turned to carving as a pastime, and they scrounged wood scraps and tree roots to make figurines, chess pieces, and caged balls and other trinkets which they sold to guards and civilians. Union and Confederate troops also carved while in their camps; anecdotes describe troops bivouacked on opposite sides of a river using small carved boats to trade coffee and tobacco.

Gunshot wounds caused severe limb injuries and resulted in 60,000 amputations among Confederate and Union troops, more such surgeries than in any other war involving American soldiers. Army surgeons often could not repair the tissue damage caused by the minié balls shot from military muskets, which shattered arm and leg bones on impact and contaminated wounds with dirt and fibers. The torn flesh became a breeding ground for the soil bacterium that causes gas gangrene (*Clostridium perfringens*); such infections were often fatal, leaving limb amputation as the only medical option. The resulting stump was often scarred or poorly healed, which some Union veterans left exposed as a personal critique of Confederate secession.[28]

Prosthetic limbs were improvised from oak staves repurposed from barrels, and peg legs were common during the war. Assuring men that families would accept their amputations became a job of army nurses; Louisa May Alcott comforted one young soldier with the promise that "she would admire the honorable scar as a lasting proof that he had faced the enemy, for all women thought a wound the best decoration a brave soldier could wear."[29] After the war, designs proliferated for artificial limbs for the thousands of surviving amputees. Government initiatives encouraged the invention of prosthetic devices of various designs, and most were carved from wood.

In Washington, the Patent Office and Office of the Surgeon General were inundated with prototypes, diagrams, and early photographs as inventors sought funding to develop prosthetic limbs with unique design and function. Between 1861 and 1873, there were 133 patents for new prosthetic limbs and devices for amputees, three times more than in the fifteen years predating the war. The Confederate Patent Office also registered designs for prosthetic legs, including the "Artificial Leg" (Confederate Patent 155, March 23, 1863) assembled by J.E. Hanger, a Confederate veteran and amputee. Using oak barrel staves, he built a leg that was hinged at the knee and ankle, which the Association for the Relief of Maimed Soldiers commissioned in large numbers from Hanger's factory in Staunton, Virginia.

The pre-war Jewett leg was also crafted from wood, but the patent does not specify the specific type, perhaps to safeguard proprietary information in an era when patent rights were often ignored. The attached linen strings simulated the action of tendons, and the toes were detachable as a single carved piece that could be replaced without damage to the rest of the device.[30] Postwar prosthetic legs perfected and improved many of the design elements found in the Jewett leg, including specifications for wood that would be both resilient and easy to bear as an appendage. The Palmer leg was a lightweight, hollow prosthetic with "English willow as the material, as long experience has shown it to be admirably adapted to the purpose."[31] Other designs such as the Reinhardt leg utilized black gum (*Nyssa sylvatica*) which was known for cross-graining (spiral arrangement of wood cells) that made the wood less likely to split.[32] Some makers favored basswood (*Tilia americana*) for carving legs, and walnut (known for shock resistance) was used specifically for crafting the heels of prostheses.[33] Wooden limbs were often carved with details such as fingernails and musculature, and makers added realistic color with paint or tinted paper decoupaged to the surface.

Some patented prosthetic legs included fittings made of rubber (often referred to as "vulcanized gum") to attach flexible toe pieces that produced a more normal gait. Plaster of Paris was used to make a mold of the stump from which a receptacle for the stump was shaped using rubber or gutta percha. Prosthetic arms and legs were also fashioned from metal rods encased in cork or papier-mâché, made from layers of glue or moistened flour mixed with shredded paper. Indeed the patent registered by Louis Tassius involved as many as three botanical materials in forming the visible surface of the prosthetic: "a mass of hardened papier-mache, surrounded with a coat of gum [rubber], and then with an envelope in leather, or, if economy be an object, in linen cloth."[34] Many makers made prosthetic hands from rubber, sometimes attached to a prosthetic arm. The prosthetic arm patented by Thomas Uren had a hollow wooden or papier-mâché forearm that was attached and moved using rubber cords.[35]

Diseases and injuries resulted in 1.5 million casualties during the Civil War, and burials occurred in mass graves near battlefields, some of them so shallow that feral pigs and other animals excavated the ground. Miasma theory suggested the presence of communicable disease (see Chapter 5), and nearby civilians resorted to carrying vials of peppermint or pennyroyal oil to mask the odors.[36] In contrast to enlisted soldiers, officers' remains were more likely to be returned to their families for burial. After the battle of Cedar Mountain (1862), the bodies of Union officers were packed in charcoal and shipped to Washington, where they were placed in metal coffins for shipping home. Charcoal was also used internally in the embalming process, a technique resembling some burial practices in ancient Egypt.[37]

Graves were marked with crosses fashioned from split rail fences or wooden panels salvaged from ammunition crates and hardtack boxes, often placed on a nearby tree. Civil War casualties led to an uptick in the mass production of coffins; some burials may have used hollow logs in place of a purpose-built coffins, but even a simple pine coffin provided some dignity in death.[38] The preference of survivors was for coffins that resisted decay, and woods were selected with this in mind; the most desirable woods were known to resist soil fungi. Chestnut wood resisted the elements when it was used in construction and fencing; the boards and posts retained their strength for years, even if they were buried or became infested with death watch or powder post beetles. Chestnut trees have minimal sapwood (the outer layer of water conducting xylem), and the heartwood (non-functioning xylem in the center of the tree) has a high tannin content. This chemistry is a factor in resisting wood fungi, although ironically, most American chestnuts eventually succumbed to the fungal disease known as chestnut blight.

The wood of American walnuts had a similar high tannin content and was also used for decay-resistant coffins. Timber cut from old growth bald cypress trees resisted fungi and insects, traits attributable to terpenes present in the bark and heartwood. The unique sesquiterpene known as cypressene accumulates over time, making bald cypress trees increasingly resistant to fungal and termite attack as they age. During the Civil War much of the bald cypress timber supply was used for shipbuilding, and competition for available timber may have made cypress coffins difficult to obtain.

Timber Products and Naval Stores

Several botanical products supplied military needs on land and water. Known from about fifty plant species, latex was the starting point for rubber used in making waterproof equipment. The primary nineteenth-century source was India rubber trees (*Ficus elastica*) native to rain forests in northern India and Malaya. Rubber tree saplings often began when their seeds landed and germinated on mossy branches and then grew prop roots extending down to the soil; wild trees grew to enormous sizes although they began as epiphytes. Latex was tapped from the woody stems and buttressed roots, yielding a substance long valued for its elasticity, adhesion, and impermeability. Chemically liquid latex is an emulsion of long-chain hydrocarbons, various resins, terpenes, and alkaloids; in plant tissue, it occurs in the anatomical canals known as laticifers and probably evolved as a deterrent to hungry herbivores.

Known as gum elastic, natural latex from *Ficus elastica* was used to make the industrial product known as India rubber, known for its flexible and waterproofing properties. However, items made from India rubber liquefied when stored at hot temperatures, a problem that intrigued Charles Goodyear who first heard of latex during the 1830s. Goodyear was a manufacturer and seller of agricultural implements, but he tinkered obsessively with latex chemistry and rubber production. Experimentation and persistence resulted in a chemical solution to rubber instability; using the serendipitous combination of latex, sulfur, and heat, Goodyear stumbled upon a chemical means for stabilizing rubber into a substance tolerant of temperature extremes. In 1844 he patented the method of vulcanization for making cloth impregnated with India rubber, and a year later established the Naugatuck India-Rubber Company in Connecticut.

Goodyear's process rendered latex suitable for use in various temperatures and climates, leading to the manufacture of gum blankets, the standard Civil War rubberized fabric used both as ground cloths and suspended for shelter (see Chapter 7). He detailed the process and its potential in two self-published volumes, *Gum Elastic and Its Varieties with a Detailed Account of Its Applications and Uses* (1853) and *The Applications and Uses of Vulcanized Gum-Elastic, with Descriptions and Directions for Manufacturing Purposes* (1853). He imagined every possible use of vulcanized latex in outfitting the military, including blankets, musket covers, powder bags, cannon covers, cartridge boxes, provision bags, bread bags, bandages, belts, canteens, knapsacks, haversacks, sails, tarpaulins, wagon floats, rafts, boats, and buoys of various types.[39] Gum blankets became known as the single most versatile and essential item made from vulcanized rubber, described by Goodyear before the Civil War "as an invaluable article for the soldier, or others who are obliged to sleep on the ground, or are in any way exposed to storms."[40] Several companies used vulcanization in manufacturing numerous rubber or rubber-impregnated products. The August 1861 price list from the Union India Rubber Company of New York described the company as the "Sole Manufacturers of the following India Rubber Goods, under Chas. Goodyear's Patents" and included tarpaulins, wagon covers, knapsacks, haversacks, canteens, navy bags, camp blankets, tents, portable boats, and water bags among the offerings.[41]

Gutta percha is a unique latex harvested from Malayan trees (*Palaquium percha*) related to chicle, and at the time of the Civil War, it had only been known in Europe and America for two decades. The trees grow slowly and must be felled in order to extract the latex, which forms in sacs in the leaves, cortex, phloem, and pith. Gutta percha is thermoplastic, meaning that it can be heated and molded, and then it solidifies when cool. Some uses resembled those of rubber, and the two were often comingled despite the thermoplastic properties of gutta percha. Wartime supplies of gutta percha were imported from Sumatra and Borneo, and the latex was used to waterproof blankets, ponchos, and clothing and to mold buttons, buckets, canteens, bayonet guards, and medical syringes (see Chapter 5). Goodyear was aware of gutta percha chemistry and observed, "Its plasticity when warmed to a temperature of about 145° Fahrenheit renders it exceedingly easy of manufacture. It retains its shape and is sufficiently hard for use at all temperatures below about 90° Fahrenheit."[42] He realized the potential of gutta percha as a substitute for whalebone and ivory, which were sometimes difficult to obtain. Eventually the most widespread use of gutta percha was in insulating the submarine telegraph cables that became widespread during the late 1800s.

The South was the primary nineteenth-century source of naval stores, the collective term for products used in shipbuilding and maintenance that were made from resin and wood of the longleaf pine (*Pinus palutris*). Southern pine forests originally colonized 130 million acres, extending along the coast from Virginia to Texas.

Periodic forest fires cleared native pine forests of scrub vegetation, leaving mature, fire-resistant longleaf pines to reach considerable size. These trees supported the naval stores industry that dated from colonial times; turpentine, rosin, tar, and pitch were produced in North Carolina and other states using pine resin as a natural resource. In *A Journey in the Seaboard Slave States*, Frederick Law Olmsted documented the antebellum naval stores industry, beginning with the work of enslaved people who incised tree trunks to start the flow of resin, which was collected in barrels made from pine staves bound with split oak.[43]

Malayan trees (*Palaquium percha*) are the source of gutta percha, a thermoplastic resin used to mold buttons, buckets, canteens, bayonet guards, and medical syringes. Gutta percha was later used to insulate the submarine telegraph cables that became widespread during the late 1800s (author's collection).

Spirits of turpentine was made by distilling pine resin in copper stills; the liquid resin was also known as crude turpentine and was harvested from living trees. According to Olmsted, "The stills used for making spirits or oil of turpentine from the crude gum, are of copper, not materially different in form from common ardent-spirit stills, and have a capacity of from five to twenty barrels."[44] Barrels of resin were converted on a vast scale into spirits of turpentine, which was used as solvent in making rubber goods, medicines, and paints. As whale oil supplies diminished, a mixture of spirits of turpentine and alcohol known as camphene became a popular lamp oil.[45] As its name implies, the mixture included camphor oil from camphor trees (see Chapter 5), but the combination of ingredients was dangerously explosive. Special camphene lamps were designed with double wicks positioned at an angle, which supposedly decreased the danger of an explosion.

Rosin was the solid substance that remained after turpentine distillation, and it was used to waterproof leather, added to lye soaps, or used as another source of oils for lamps or lubrication. Olmsted noted that the amount of rosin produced as a by-product outstripped demand and observed that at one distillery thousands of gallons were discarded onto the ground "glistening like polished porphyry," reminiscent of an igneous rock with a "deep purple-red tone."[46] Tar was the product of slowly burning pine heartwood with minimal oxygen, a process that yielded a mixture of aromatic hydrocarbons. Burning was conducted in a tar kiln, layers of pine wood insulated with branches and soil, and a drain (often a hollow log); barrels were positioned to collect tar as it dripped from the burning logs. The process required precision for success, as Olmsted observed, "The proper burning of the kiln to produce the most tar, is an art to be learned by practice."[47] Pine tar was used to preserve wood and weatherproof ropes and rigging, and pitch (a mixture of tar and spirits of turpentine boiled into a semi-solid) was used to coat the hulls of wooden vessels to make them watertight. Rope soaked in pitch was known as oakum, and it was driven into the cracks between planks to make watertight seals. Farmers were familiar with the apparent antimicrobial properties of tar which prevented sepsis in livestock wounds and protected fence posts from fungal decay.[48]

In the antebellum years, most naval stores were shipped to ports in New York and Europe, but the pine-based industry was drastically disrupted by Civil War blockades. Small producers could not trade their naval stores for flour and other necessities, and early in the war summertime droughts caused their corn and potato crops to fail, bringing some families in pine regions of the Confederacy to near starvation. Union states sought wartime alternatives to southern naval stores, and in the years after the war, kerosene replaced camphene as a cheap lamp oil.

Military Engineering and Construction

Trees provided timber for constructing military ships, buildings, bridges, roads, and fortifications. The properties of woods were gleaned from observation and confirmed through trial and error, but army engineers were limited by the diversity of local forests and availability of commercial lumber in demand for other wartime uses. In contrast, shipbuilding required wood that withstood compression, shearing, and moisture. Union and Confederate navies used oak for constructing the frames of their ships, and although the trees are small to medium in stature, the wood of red mulberry was also used for ship framing and flooring, and the fibrous inner bark was used for cordage. The shock-resistant wood of elm (*Ulmus americana*) was used to craft the block-and-tackle rigging used to hoist sails and unload cargo. Other naval timber included planks cut from old growth bald cypress trees, wood known as decay-resistant but also in demand for railroad ties. Native to the South, black locust (*Robinia pseudoacacia*) had similar uses in constructing ships, railroad tracks, and fencing; indeed its widespread planting for railroad expansion led to invasive growth outside of its native range. General Sherman directed federal troops to build fires from wooden railroad ties as troops moved through Confederate states; rails were heated on the fire and bent into unusable shapes that became known as "Sherman's neckties."

Trestle bridges were constructed from logs, timber cut by army sawmills, and boards that were scavenged from farms and homes, following the detailed instructions in military manuals for constructing bridges across ravines.[49] Water posed challenges with regard to flotation and current; Union regiments devised a bridge across the Mississippi using cotton bales for flotation and boards scavenged from cotton warehouses.[50] However, near disaster occurred when Confederate engineers repurposed ammunition crates to anchor a pontoon bridge across the Potomac following General Lee's defeat at Gettysburg. The rock-filled (but unanchored) wooden crates shifted on the muddy river bottom, causing the bridge to separate into sections.[51]

Knowledge of pontoon bridges was fundamental to military engineering, and part one of J.C. Duane's *Manual for Engineer Troops* was titled "Ponton Drill" and covered in detail the transport, structure, precise measurements, and assembly of portable floating bridges.[52] Purpose-built to move troops, wagons, and horses across rivers, pontoon bridges were buoyed by air-filled cylindrical rubber bags, wooden bateaux (shallow boats), or lightweight canvas-covered wooden frames. However, officers soon learned that the rhythmic force of hundreds of men marching in unison could defeat buoyancy, and so they instructed troops to stride normally rather than to march in step. Sections of a pontoon bridge were held together by long wooden boards (balks) on which the flooring (chess) was oriented at a right angle; expert lashing with rope held everything in place.

Success of a pontoon bridge pivoted on the buoyancy of the bateaux or other floats, and no single design was ideal. Bateaux required seasoned wood or they warped and leaked, and canvas floats required maintenance. The latter were easier to move, but enzymes produced by mildew fungi weakened the canvas floats, which then needed frequent repair. Known as Cumberland pontoons, portable bridges with wooden frames and canvas coverings were used by General Sherman in the final campaigns of the Civil War.

Floating portable pontoon bridges allowed troops, wagons, and horses to cross rivers. Wooden floats required seasoned wood or they warped and leaked; sections of the bridge were held together by long wooden boards, and expert lashing with rope held everything in place (*Harper's Weekly*, 1863).

Army engineers used logs cut from nearby forests and woodlots for constructing corduroy roads, bridges, rafts, and blockhouses (used for shelter against enemy fire) and for stabilizing the steep slopes of trenches. Corduroy roads were used for centuries in traversing wet areas, and they were essential in alluvial zones known for mud and quicksand. The standard technique involved felling trees, trimming them into logs, and laying them parallel with the roadsides. A perpendicular layer of shorter logs was attached at the sides to elevate the road surface above the mud. Sometimes construction required multiple layers of logs before the road could support military traffic without sinking into the ground. Scavenged fence rails and scrap wood were often incorporated into corduroy roads, which sometimes extended for a mile or more through wetlands or along the access routes to river banks.

General Grant's memoirs included several references to corduroy roads in navigating mud. In March 1865, Union troops near Petersburg, Virginia, were virtually trapped by quicksand, and roads made from trees offered the only solution: "It became necessary therefore to build corduroy roads every foot of the way as we advanced, to move our artillery upon. The army had become so accustomed to this kind of work, and were so well prepared for it, that it was done very rapidly."[53]

Some roads were corduroyed in anticipation of spring rain and quagmire that might result from the movement of troops along dirt roads. The speed and skill of Union engineers in building corduroy roads was a key factor in winning the war, as noted by Confederate General P.T.G. Beauregard in 1865: "When I learned that Sherman's army was marching through the Salk swamps, making its own corduroy roads at the rate of a dozen miles a day and more, I made up my mind that there had been no such army in existence since the days of Julius Caesar."[54]

Most corduroy roads during the Civil War were built in haste as temporary structures from any available wood. However, some were intended as permanent structures, and some were built as raised roads supported by pilings, in which case wood selection would have been important to avoid deterioration and decay. Chestnut was superior to pine in this regard, but engineers typically used trees that grew in the immediate area to avoid hauling logs any distance. Drainage ditches were sometimes excavated under the roads; working in mud or standing water, engineers replaced shifting logs that posed dangers to horses and wagons.[55] There is no doubt wood properties posed additional problems; softwoods (pines and other conifers) lack the flexible, thick-walled fibers that strengthen hardwoods, and thus conifer timber was generally less dense and more prone to the compression and splintering that could be caused by horses and wagons.

Many Civil War fortifications relied on an internal framework constructed from substructures made of timber and brushwood. Military manuals provided detailed instructions for crafting the gabions, fascines, sap rollers, hurdles, and other substructures essential to siege warfare, in which fortifications held the enemy at bay. Army engineers were limited to the botanical materials available in the theater of war, and they were trained to follow engineering instructions as closely and precisely as possible under field conditions. Gabions were essentially open cylindrical baskets in which brushwood and woody grapevines were woven around wooden pickets, with a finished diameter of two feet and height of 2.75 feet. They were prepared in advance and stockpiled until needed for construction. Once positioned, gabions were filled with earth; their strength came from the tons of dense soil and rocks that a fortification made of gabions could contain.

Civil War army engineering companies built fortifications known as traverses,

Gabions were open baskets woven from brushwood and grapevines, made in advance and stockpiled for building fortifications. Gabions were positioned and then filled with earth; their strength came from the tons of dense soil and rocks that a fortification made of gabions could contain (*Harper's Weekly*, 1863).

and the strongest of these began with a lower tier of gabions laid out in a rectangle approximately 12 feet deep by 24 feet wide. Additional gabions were placed inside the rectangular area, and the entire site was then filled with earth. A top tier of more gabions layered with bundles of brushwood (known as fascines; see below) completed the structure. Willow branches (known as withes) may have been used to bind the gabions together into a cohesive structure that was able to withstand artillery fire. Depending on their design and density, traverses offered protection from hostile fire that ranged from shell fragments to cannonballs.

Fascines were devised from brushwood tied into bundles of standard width and length. Collected from felled trees, twigs were sorted by size, and branches less than two inches in diameter were used for the bundles. According to *A Treatise on Field Fortification* (1862) by D.H Mahan, "To make a fascine, straight twigs are selected, between the thickness of the little finger and thumb, the longer the better; they should be stripped of the smaller twigs ... a fascine horse, is put up, by driving two stout poles obliquely into the ground ... this forms the horse, on which the twigs are laid to be bound together."[56] The bundles of brushwood were bound together tightly with bands made of flexible willow stems. In addition to their use in construction, fascines were used in the revetments that stabilized the sloping walls of fortifications. Rows of the bundles were partially buried in soil and held in place with pickets pounded into the earth. Squares of sod stabilized the earthworks against movement, based on the strength of intertwined roots which held the soil in place; as Mahan suggested, "The sods should be cut from a well-clothed sward, with the grass of a fine short blade, and thickly matted roots."[57]

Like gabions, hurdles were based on the idea of wickerwork, and their construction began with ten pickets driven into the ground in the shape of an arc eight feet in length. The completed hurdle consisted of a curved wall of woven branches with a typical height of six feet.[58] They were used in revetments and in lining *trous de loup*, the conical pits excavated as defensive obstacles against the movement of enemy troops. Such pits were typically over six feet in depth, and each had a sharpened wooden stake at the base, perhaps masked by a hurdle. They were excavated singly or in clusters, usually in the area approaching a defended site.

Entire trees served as obstacles to troops on the march. Several trees were felled adjacent to a ditch, and their trunks were partially buried to make an arrangement known as an abattis. Military manuals advised felling the trees in the direction of enemy movement and weaving the branches together into a head-on wall of dense vegetation. Smaller branches were then systematically trimmed and sharpened.[59] Approaching troops found the tangled branches from multiple trees difficult or impossible to penetrate, especially when the abattis was made from a thorny tree such as Osage orange (see Chapter 2). A similar strategy was used in the defensive obstacles known as *chevaux de fries* (horses of Holland, where they were originally used, also known as *fraise*). Small tree trunks were sharpened and then oriented and anchored at angles to prevent the movement of cavalry troops, resembling the sharpened wood stakes (known as *tetes du pont*) that were used to block access to bridges.

Like gabions, the structures known as sap-rollers began as woven brushwood cylinders, but they were made in pairs so that one fit inside the other, leaving an eight-inch gap to pack tightly with fascines, larger woody branches, or pine stakes. The completed cylinder was bullet-proof and could be positioned to protect troops tasked with digging the zig-zag trenches known as saps.[60] Some engineers used the native bamboo known as cane (*Arundinaria* spp.; see Chapter 3) to weave and fill sap-rollers; as an abundant natural resource, cane was also used in making fascines to place over trenches for cover and some protection against musket balls.[61] Botanically speaking, the stems are not actually wood because bamboo and other grasses do not form a vascular cambium, but the shoots are strong and flexible as a result of elongated cells (sclerenchyma tissue, characterized by cell walls reinforced by additional cellulose and lignin) which accompany the bundles of conducting tissue.

Loss and Gain

Despite the economic and social importance of agriculture, in 1860 most of the land (probably eighty percent) in Confederate states was uncultivated, providing habitats for native and naturalized plants to colonize and reproduce. Cattle roamed freely in fields and forests, and pigs foraged for mast, the collective windfall of acorns and other nuts. However, forests of all types and ages burned during the war, the result of artillery explosions, unchecked camp fires, and arson. Pine forests were particularly vulnerable to fire because the terpene-rich pine resin is highly flammable, and some fires were intentionally set at the incisions made for collecting resin.[62]

Forests, fields, farms, gardens, and roads across the South bore the scars of military activity. The destruction of timber for military construction and campfires consumed hundreds of acres of timber. A mile of corduroy road alone required roughly 200 mature trees, depending on their size, which resulted in the clear-cutting of three or four acres of mature woodland. Saplings and small trees were literally shredded by musket fire, but some mature trees survived both lead bullets and extreme heat. In many cases, the only survivors were large trees that bore charred scars from flames and gunshots; bullets and cannonballs became embedded in wood as trunks continued to grow in diameter. In a landscape pock-marked by damaged trees, soldiers often collected pieces of wood or bark as mementoes of battles or surrenders.

Trees were intentionally felled to impede troop movements and to excavate canals that provided access to strategic islands in the Mississippi River.[63] In some cases this involved cutting through old growth bald cypress swamps, dense habitats in which woody "knees" grew from submerged roots. Union soldiers and freedmen struggled to remove trees and extensive roots before excavating waterways with shovels. Although virgin forest still covered some areas of the South, historical status was not a factor in deciding which trees to cut to meet wartime timber demand. Virgin walnut, chestnut, and bald cypress were harvested from ancient primary forests for wartime use. The Department of Agriculture observed the damaged wartime landscape, but placed blame squarely with the Confederate support of slavery and emphasis on cash crops: "In most of the counties throughout the tide-water region of Virginia there was, before the present war, a great extent of waste land that had been impoverished by injudicious cultivation in corn and tobacco.... A large portion of the land in eastern Virginia now lies desolate, having been brought to that condition by the scourge of civil war, resulting from disobedience to the Divine law and disregard to the first principles of civil liberty."[64]

The federal government envisioned the restoration of southern forests as part of postwar Reconstruction. The Department of Agriculture summarized the effect of war, "The destruction of forests and timber during the war of the rebellion has been immense. Both armies, the Union and the rebel, have destroyed it. Much has been ruined by accidents; it has been removed for military purposes, both by the axe and fire; it has been taken to supply fuel for the armies, to erect fortifications, to hinder the movements of the enemy, and to open the country for military movements.... The general government, in its grand and sudden expansion of our navy, has almost stripped some of the best sections of the whole country of its very best timber—the white oak—which has gone to the navy yards and contractors' docks in untold quantities."[65] The proposed solution was innovative for the time, "Let extensive, protracted and scientific experiments in the propagation and cultivation of forest trees be established.... Our country, in the general excellence and variety of its timber, exceeds Europe and demands that we should study and learn for ourselves what our country can do for its native trees, and what our trees can do for our country...this subject should receive the immediate attention of our government, and enjoy its fostering care."[66] However, the USDA did not initiate a Division of Forestry until 1881.

The regrowth of forests occurred naturally, even though land was still cleared for agriculture and railroads after the war. Fortifications, defensive obstacles, and corduroy roads were made of wood and other botanical materials, which quickly decayed in Southern climates, and the landscape began to revert to its natural contours. Perhaps ironically, wartime fires improved soil fertility by returning mineral salts to the forest floor. Across the South, trees reestablished in abandoned agricultural fields and plundered forests; woodlands grew quickly from seeds dispersed by wind and animals and from surviving root systems that sprouted new shoots. Damaged trees regenerated from epicormic buds that lie dormant just under the bark; fire and increased exposure to light often stimulates bud growth and thus helped to restore tree crowns and the forest canopy. However, saplings in former battlefields may also have been affected by the production of lead oxides as damaged wood decayed and released lead bullets into the soil. Lead oxides inhibit the growth of soil microbes which convert organic matter into fertile topsoil layers, and the compounds may also affect photosynthesis by inhibiting the uptake of magnesium, an element essential to chlorophyll production. At the same time, non-native trees escaped cultivation from groves and gardens, quickly naturalizing in disturbed sites. The synergistic combination of a warm climate, high rainfall, and long growing season (the selfsame conditions that encouraged cash crops) promoted timber growth and the natural restoration of Southern forests.

Afterword

I have relied on wisdom from ethnobotany, economic botany, and plant science to examine the role of plants during the Civil War—in agriculture, diet, medicine, gardening, timber, landscapes, military engineering, and numerous other ways. The core issue of the Civil War was botanical, originating with colonial crops that included tobacco and indigo. Cash crops became the basis of the Southern economy, an agricultural system that could only be sustained with cheap labor. Once cotton gins solved the problem of fiber processing, cotton cultivation soared, resulting in the King Cotton economy that drove the antebellum South. Southerners demanded the right to operate on a vast plantation scale, and some tried to justify enslavement as part of the natural order. Natural resources were exploited as primary forests were clear-cut, soils became nutrient-depleted, and invasive plants escaped cultivation. However, enslavement for agricultural profit was the core issue in secession that led to war. The botanical landscape was merely collateral damage.

The planter class sought European trappings, but ornate furnishings and pleasure gardens did nothing to assuage malaria and sweltering heat. Antebellum houses and land were neglected as the plantation system collapsed, and social decay was linked to environmental and agricultural ruin. Farm machinery, soil amendments, and seed stock were all in short supply, and troops on the march burned timber fences and made off with stored crops. As the Civil War unfolded, many were forced to return to old survival strategies that required practical botanical knowledge. Confederate civilians turned to subsistence gardens, staple crops, medicinal herbs, and homespun cloth, skills acquired by trial and error and practiced well into the Reconstruction era. Cultivated fields and forests sustained damage from fortifications and battles, which took years to diminish and heal.

Confined to plantation landscapes, enslaved people had brush arbors and vegetable gardens, outdoor spaces where African traditions were practiced and oral traditions shared. Brush arbors in particular provided safe spaces for private thoughts and brief respite from misery. Foods with African origins became part of American culinary history, and enslaved people practiced herbal medicine, including contraception to limit the births of enslaved children; the irony that cotton was used as an abortifacient cannot be overlooked. The expertise of enslaved people and freedmen saved lives when the only available medicines came from local flora and the only available foods came from subsistence gardens.

Soldiers also became practical botanists. They collected plants for medicines, emergency foods, and building materials for shelters and fortifications. Drug shortages caused by the Union blockade resulted in the need for quinine and opium substitutes, and experimentation with other medicinal species was widely encouraged. Woody plants also provided the raw material for siege warfare; fortifications and corduroy roads were constructed from trees, shrubs, and vines. Awareness of plant structure and strength were important in sourcing materials from nature. Stringencies forced innovation, especially in the South, and wild plants were diverse and abundant. Hands-on tinkering revealed many new and alternative uses for native and non-native plants.

Reconstruction (1865–1877) resulted in fundamental changes to Southern agriculture. Many old fields had reverted to wild tangles of briars and broomsedge, and like the South generally, they required care and reclamation. Tenant farming expanded as landowners subdivided large plantations into smaller acreage that was rented to freedmen and indigent farmers. Sharecropping may have originated in Mississippi, but the practice became widespread as a system in which landowners and farmers shared the harvested crops. Small-scale farming became a way of life for many, but soils were often depleted, and agricultural science was unknown.

The Morrill Land-Grant Act of 1862 was extended to former Confederate states, leading to land-grant agricultural colleges in which progressive practices were developed on model farms and shared through extension services. The Morrill Act of 1890 funded historically Black colleges and universities at which the descendants of freedmen studied farming and agriculture, but the effects caused by antebellum plantation agriculture persisted. Born at the end of the Civil War, George Washington Carver investigated the effects of years of cotton cultivation on southern soils. Working at the private Tuskegee Institute, he promoted the cultivation of high-nutrient crops such as peanuts and sweet potatoes, some of the same crops that enslaved workers cultivated in their plantation gardens.

Botanical history illuminates the conflict between the North and South, wartime stringencies and solutions, and the overarching importance of plants in daily lives during the Civil War, which was followed by years of regeneration—natural, agricultural, and social. Speaking at Yale on July 26, 1865, the Rev. Horace Bushnell drew a parallel between the Civil War and botanical regeneration after the Noachian flood, observing that in the aftermath of destruction "the seeds of a true public life are in the soil, waiting to grow apace. It will be as when the flood of Noah receded ... no sooner were the waters down, than the oaks, and palms, and all the great trees, sprung into life, under the dead old trunks of the forest, and the green world reappeared even greener than before."[1]

Chapter Notes

Chapter 1

1. R.R. Menard, "How Sugar and Tobacco Planters Built Their Industries and Raised an Empire," *Agricultural History* 81, No. 3, Summer 2007, pp. 309–332, p. 318.
2. Federal Writers' Project, *American Slave Interviews: Alabama Narratives*, Vol. 1 (Washington, D.C., Works Progress Administration, 1941), p. 95 [Katherine Eppes].
3. P.S. Taylor, "Plantation Laborer Before the Civil War," *Agricultural History* 28, No. 1, Jan. 1954, p. 111.
4. *Ibid.*
5. Henry William Ravenel, *Private Journal, 1865–1866*, Sept. 1865, p. 59, University of South Carolina University Libraries, https://digital.tcl.sc.edu/digital/collection/rav/id/5491/rec/3.
6. C. Darwin, *Journal of Researches into the Natural History and Geology of the Countries Visited During the Voyage of H.M.S. Beagle Round the World, Under the Command of Capt. Fitz Roy, R.A.*, ed. 2 (London, John Murray, 1845), pp. 499–500.
7. Letter from C. Darwin to A. Gray, April 19, 1865, https://www.darwinproject.ac.uk.
8. A. Gray, *A Manual of the Botany of the Northern United States: From New England to Wisconsin and South to Ohio and Pennsylvania Inclusive (the Mosses and Liverworts by Wm. S. Sullivant,) Arranged According to the Natural System* (Boston, James Munroe, 1848).
9. A. Gray, *Manual of the Botany of the Northern United States, Including the District East of the Mississippi and North of North Carolina and Tennessee*, ed. 5 (New York, Ivison, Taylor, Blakeman, and Company, 1867).

Chapter 2

1. A.J. Downing, *A Treatise on the Theory and Practice of Landscape Gardening* (New York, Wiley and Putnam, 1841), pp. 23–24.
2. A.J. Downing, *Rural Essays* (New York, Leavitt and Allen, 1857), p. 405.
3. "An Essay," *The American Cotton Planter* 1, Sept. 1853, p. 269.
4. F. L. Olmsted, *A Journey in the Seaboard Slave States; with Remarks on Their Economy* (London, Sampson Low, Son, and Company, 1856), p. 417.
5. F.P. Porcher, *Resources of the Southern Fields and Forests* (Richmond, Evans and Cogswell, 1863), p. 524.
6. A.J. Downing, *A Treatise on the Theory and Practice of Landscape Gardening*, p. 121.
7. Porcher, p. 236.
8. P.V. Petrakis, K. Spanos, A. Feest, and E. Daskalakou, "Phenols in Leaves and Bark of *Fagus sylvatica* as Determinants of Insect Occurrences," *International Journal of Molecular Sciences* 12, 2011, p. 2777.
9. "Southern Native Trees and Shrubs," *The American Cotton Planter* 1, July 1853, p. 223.
10. "Report and Review," *American Cotton Planter* 2, No. 3, March 1854, pp. 234–235.
11. Letter to Thomas Jefferson from William Bartram, Oct. 29, 1808, National Archives Founders Online, https://founders.archives.gov.
12. "An Essay," *The American Cotton Planter* 1, No. 9, Sept. 1853, p. 270.
13. F.L. Olmsted, *A Journey in the Seaboard Slave States; with Remarks on Their Economy*, pp. 416, 693.
14. *Ibid.*, p. 416.
15. *Ibid.*, p. 566.
16. *Ibid.*, pp. 89–90.
17. A.J. Downing, *A Treatise on the Theory and Practice of Landscape Gardening*, p. 204.
18. J.A. Warder, *Hedges and Evergreens* (New York, O.A. Moore, 1859), p. 275.
19. *Ibid.*, pp. 43, 274.
20. *Ibid.*, pp. 24–25.
21. F.L. Olmsted, *The Cotton Kingdom: A Traveller's Observations on Cotton and Slavery in the American Slave States* (New York, Mason Brothers, 1862), p. 282.
22. F.L. Olmsted, *A Journey in the Back Country* (New York, Mason Brothers, 1860), p. 13.
23. "The Willow and Osage Orange," *The American Cotton Planter* 3, No. 5, May 1855, p. 148.
24. "Premium List of the First Annual Fair of the Alabama State Agricultural Society," *The American Cotton Planter* 3, No. 6, June 1855, p. 173.
25. "Various Inquiries Answered," *The American*

Cotton Planter and Soil of the South 3, No. 11, Nov. 1859, p. 338.

26. C. Barlow, "Anachronistic Fruits," *Arnoldia* 61, No. 2, 2001, pp. 16–18.

27. E.F. Frazier, "The Negro Slave Family," *The Journal of Negro History* 15, No. 2, April 1930, p. 244.

28. J.D. Cornelius, *Slave Missions and the Black Church in the Antebellum South* (Columbia, University of South Carolina Press, 1999), pp. 8–10.

29. L.M. Brady, *War Upon the Land: Military Strategy and the Transformation of Southern Landscapes During the American Civil War* (Athens, University of Georgia Press, 2012), pp. 22, 55–56.

Chapter 3

1. P.W. Gates, *Agriculture and the Civil War* (New York, Alfred A. Knopf, 1965), p. 371.

2. D.G. Faust, "The Rhetoric and Ritual of Agriculture in Antebellum South Carolina," *The Journal of Southern History* 45, No. 4, Nov. 1979, p. 560.

3. *Ibid.*, p. 564.

4. *Ibid.*, p. 547.

5. F.P. Porcher, *Resources of the Southern Fields and Forests* (Richmond, Evans and Cogswell, 1863), p. 249.

6. "Rotation for a Small Farm," *The Country Gentleman* 15, No. 1, Jan. 5, 1860, p. 11.

7. *Ibid.*

8. R.D. Powell, "Agriculture," *The American Cotton Planter* 1, No. 3 n.s., March 1857, pp. 326-328.

9. "Corn Culture—Prize Essay," *Southern Cultivator* 19, No. 3, March 1861, pp. 75–76.

10. "Supplying Plant Food at the Surface," *The Country Gentleman* 15, No. 2, Jan. 12, 1860, p. 27.

11. "Gradual Deterioration of the Soil," *The American Cotton Planter* 4, No. 7 n.s., July 1860, p. 304.

12. "Virginia: Her Past, Present, and Future," *Report of the Commissioner of Agriculture for the Year 1864* (Washington, D.C., Government Printing Office, 1865), p. 20.

13. "Agricultural Science and Literature," *The American Cotton Planter* 4, No. 3 n.s., March 1860, p. 107.

14. "Report," *Report of the Commissioner of Agriculture for the Year 1862* (Washington, D.C., Government Printing Office, 1863), p. 19.

15. "Agriculture in Schools," *The Country Gentleman* 15, No. 6, Feb. 9, 1860, p. 99.

16. F.L. Olmsted, *A Journey in the Seaboard Slave States; with Remarks on Their Economy* (London, Sampson Low, Son, and Company, 1856), p. 65.

17. E.L. Rice, "Allelopathic Effects of *Andropogon virginicus* and Its Persistence in Old Fields," *American Journal of Botany* 59, No. 7, pp. 752–755.

18. M.A. Stewart, "From King Cane to King Cotton: Razing Cane in the Old South," *Environmental History* 12, No. 1, Jan. 2007, pp. 68–71.

19. Ibid., pp. 61–62.

20. F.L. Olmsted, *A Journey in the Seaboard Slave States; with Remarks on Their Economy*, p. 241.

21. "Marl," *Southern Cultivator* 19, No. 5, May 1861, p. 160.

22. U.S. Centennial Commission, *International Exhibition, 1876 [Reports]*, vol. 9, 1876, pp. 353–55.

23. "Lime as a Manure," *Southern Cultivator* 19, No. 2, Feb. 1861, pp. 49–50.

24. "Saving Manure," *Southern Cultivator* 19, No. 7, July 1861, p. 231.

25. M. Filmore, "First Annual Message," Dec. 2, 1850, The Presidency Project, www.presidency.ucsb.edu.

26. "Guano," *Southern Field and Fireside* 2, No. 16, Sept. 6, 1860, p. 126.

27. "Prize Essay on Commercial Manures," *Southern Cultivator* 19, No. 2, Feb. 1861, pp. 42–43.

28. F.L. Olmsted, *A Journey in the Seaboard Slave States; with Remarks on Their Economy*, p. 303.

29. "Artificial Guano. A Cheap and Good Fertilizer," *Southern Cultivator* 19, No. 1, Jan. 1861, pp. 19–20.

30. "Leaf Mould—How to Prepare," Southern Cultivator 19, No. 1, Jan. 1861, p. 13.

31. Porcher, p. 504.

32. *Ibid.*, p. 178.

33. *Ibid.*

34. F.L. Olmsted, *A Journey in the Seaboard Slave States; with Remarks on Their Economy*, p. 466.

35. *Ibid.*, p. 418.

36. "Shall We Raise Tobacco?" *The New England Farmer* 15, No. 12, Dec. 1863, pp. 377–378.

37. *Ibid.*, p. 378.

38. *Ibid.*

39. "Shall We Raise Tobacco?" *The New England Farmer* 16, No. 1, Jan. 1864, pp. 24–25.

40. R.R. Harden, "The Rot in Cotton," *The Farmers' Register* 5, No. 1, May 1837, p. 24.

41. "Plantation Work for June," *The American Cotton Planter* 1, No. 3 n.s., March 1857, p. 163.

42. W.C. Bandy, "Cotton Seed Manure," *The American Cotton Planter* 1, No. 3 n.s., March 1857, p. 326.

43. "Potash—A Hint to Farmers," *Southern Cultivator* 19, No. 12, Dec. 1861, p. 305.

44. "Cotton Culture," *Southern Cultivator* 20, Nos. 11 and 12, Nov. and Dec. 1862, p. 207.

45. E. Sanderson, "Feeding Hogs on Cotton Seed," *The American Cotton Planter* 1, No. 3 n.s., March 1857, p. 315.

46. "New Gas Works," *Scientific American* 13, No. 27, March 13, 1858, p. 212.

47. "Coal Oil and Cotton Seed Oil and Cake," *The Country Gentleman* 15, No. 1, Jan. 5, 1860, p. 19.

48. "Cotton Seed Oil and Cake," *Southern Cultivator* 19, No. 1, Jan. 1861, pp. 11–12.

49. "Cotton Seed Huller," *Southern Cultivator* 17, No. 12, Dec. 1859, p. 376.

50. "Plant Corn!" *Southern Cultivator* 19, No. 3, March 1861, pp. 84–85.

51. "Bread for the South," *Southern Cultivator* 19, No. 1, Jan. 1861, p. 13.
52. "Make Provision Crops," *Southern Cultivator* 20, Nos. 11 and 12, Nov. and Dec. 1862, p. 213.
53. "Cow Peas," *Southern Cultivator* 21, Nos. 9 and 10, Sept. and Oct. 1863, p. 110.
54. "Hints for July and August," *Southern Cultivator* 20, Nos. 7 and 8, July and Aug. 1862, pp. 129–131.
55. "Culture of the Ruta Baga," *The Country Gentleman* 15, No. 12, March 22, 1860, p. 187.
56. "Hints for the Season," *Southern Cultivator*, vol. 21, nos. 7 and 8, July and Aug. 1863, p. 93.
57. "Jerusalem Artichoke," *Southern Cultivator*, vol. 22, no. 3, March 1864, p. 51.
58. "Peas Beans and Mangolds," *The Country Gentleman* 15, No. 21, May 24, 1860, p. 333.
59. "The Cultivation of Sorghum," *Southern Cultivator* 19, No. 6, June 1861, p. 197.
60. "Refuse of Sorgho," *Southern Cultivator* 21, Nos. 7 and 8, July and Aug. 1863, p. 94.
61. "Sorghum for Soiling," *The Country Gentleman* 15, No. 13, March 29, 1860, p. 205.
62. "Women and Agriculture," *Southern Cultivator* 19, No. 4, April 1861, pp. 111–112.
63. "Female Volunteers," *Southern Cultivator* 19, No. 5, May 1861, p. 154.
64. "Industry of Southern Women," *Southern Cultivator* 22, No. 1, Jan. 1864, p. 22.
65. "Spading and Subsoiling," *Southern Cultivator* 23, No. 2, Feb. 1864, p. 28.
66. "Manures," *Southern Cultivator* 23, No. 2, Feb. 1864, pp. 25–26.
67. "A Cheap and Ready Mode of Composting Night Soil," *The Country Gentleman* 16, No. 6, Feb. 9, 1860, pp. 90–91.
68. "When to Sell Hay," *American Agriculturalist* 21, No. 10, Oct. 1862, p. 301.

Chapter 4

1. "The Marblehead Drumhead Cabbages," *New England Farmer* 13, No. 4, April 1861, p. 192.
2. "Cultivation of Potatoes," *Southern Cultivator* 19, No. 3, March 1861, pp. 83–84.
3. "Gardening for Farmers," *Southern Cultivator* 19, No. 2, Feb. 1861, p. 63.
4. *Confederate Receipt Book: A Compilation of Over Hundred Receipts Adapted to the Times* (Richmond, West and Johnson, 1863), p. 5.
5. M. Randolph, *The Virginia House-Wife* [reprint of the 1824 edition] (Columbia, University of South Carolina Press, 1984), p. 154.
6. F.P. Porcher, *Resources of the Southern Fields and Forests* (Richmond, Evans and Cogswell, 1863), p. 64.
7. J. Sumner, *American Household Botany: A History of Useful Plants 1620–1900* (Portland, OR, Timber Press, 2004), p. 137.
8. "How to Raise Early Tomatoes," *Southern Cultivator* 19, No. 1, Jan. 1861, p. 30.
9. A.W. Chase, *Dr. Chase's Recipes; or Information for Everybody: An Invaluable Collection of About 800 Recipes*, 30th ed. (Ann Arbor, 1866), pp. 24, 315.
10. Randolph, pp. 201, 202, 241.
11. *Confederate Receipt Book*, p. 10.
12. C. Beecher, *Miss Beecher's Domestic Receipt-Book*, ed. 3 (New York, Harper and Brothers, 1862), p. 184.
13. "The Orchard," *The American Cotton Planter and Soil of the South* 13, No. 1, Jan. 1859, p. 26.
14. "A Lesson of the Times: The South Must Be Independent," *Southern Cultivator* 19, No. 1, Jan. 1861, p. 15.
15. "Apples," *The Ladies' Home Magazine* 15, Dec. 1860, p. 372.
16. Fruitland advertisement, *The American Cotton Planter and Soil of the* South 14, No. 2, Feb. 1860, p. 83.
17. "Fruits—Re-Production of Varieties," *Southern Cultivator* 19, No. 2, Feb. 1861, pp. 60–61.
18. "The Fruit Orchard and Nursery," *The American Cotton Planter and Soil of the South* 14, No. 2, Feb. 1860, p. 82.
19. *Confederate Receipt Book*, p. 7.
20. *Ibid.*, pp. 9–10.
21. Porcher, pp. 142–143.
22. C. Beecher and H.B. Stowe, *The American Woman's Home* (New York, J.B. Ford and Company, 1869), p. 190.
23. B.T. Allen, *Sugaring Off: The Maple Sugar Paintings of Eastman Allen* (Williamstown, MA, Clark Art Institute, 2005), p. 39.
24. "Some Outlines of the Agriculture of Maine—Maple Sugar and Sirup," *Report of the Commissioner of Agriculture for the Year 1862* (Washington, D.C., Government Printing Office, 1863), p. 54.
25. Porcher, pp. 80–83.
26. Randolph, p. 172.
27. *Ibid.*, pp. 100, 153, 154, 157–158.
28. F.L. Olmsted, *A Journey in the Seaboard Slave States; with Remarks on Their Economy* (London, Sampson Low, Son, and Company, 1856), pp. 80, 92.
29. *Confederate Receipt Book*, pp. 5–6.
30. Beecher and Stowe, p. 171.
31. Porcher, p. 549
32. Olmsted, p. 478.
33. "A Good Cheap Pie," *The American Cotton Planter and Soil of the South* 14, No. 6, June 1860, p. 263.
34. "Queen Victoria's Children," *Southern Cultivator* 19, No. 2, Feb. 1861, p. 59.
35. Porcher, p. 552.
36. "Cotton as an Antiseptic," *Southern Cultivator* 19, No. 8, Aug. 1861, p. 246.
37. Porcher, p. 64.
38. *Ibid.*, p. 383.
39. P.W. Gates, *Agriculture and the Civil War* (New York, Alfred A. Knopf, 1965), p. 80.
40. "Indian or Mixed Pickles," *Godey's Lady's Book and Magazine* 65, Nov. 1862, p. 500.
41. Beecher, pp. 160–165.
42. Porcher, p. 532.

43. *Confederate Receipt Book*, p. 14.
44. *Ibid.*, p. 8
45. Porcher, p. 472.
46. K. Ouchley, *Flora and Fauna of the Civil War: An Environmental Reference Guide* (Baton Rouge, Louisiana State University Press, 2010), pp. 25–29, 51.
47. L.M. Alcott, *Hospital Sketches* [reprint of the 1863 edition] (Boston, Applewood Books, 1986), p. 63.
48. J.M. Sanderson, *Camp Fires and Camp Cooking: Culinary Hints for the Soldier* (Washington, D.C., Government Printing Office, 1862), p. 4.
49. *Ibid.*, pp. 13–14.
50. U.S. Quartermaster General, *Bread and Bread Making* (Washington, D.C., Government Printing Office, 1864).
51. "Experiments on Alcoholic Fermentation and Its Causes," *Report of the Commissioner of Agriculture for the Year 1864* (Washington, D.C., Government Printing Office, 1865), pp. 519–532.
52. Sanderson, pp. 5–6.
53. *Ibid.*
54. F. Burr, Jr., *The Field and Garden Vegetables of America* (Boston, J.E. Tilton and Company, 1865), pp. 434–481.
55. Sanderson, pp. 6–7.
56. Porcher, p. 177.
57. *Ibid.*, pp. 405–406.
58. "Campaigning Axioms," *Southern Cultivator* 19, No. 5, May 1861, p. 194.
59. Gates, pp. 97–99.
60. E.M. Coulter, "The Movement for Agricultural Reorganization in the Cotton South during the Civil War," *Agricultural History* 1, No. 1, Jan. 1927, p. 15.
61. Porcher, pp. 143,171.
62. "Sparkling Wine—How Made!" *Southern Cultivator* 19, No. 2, Feb. 1861, p. 63.
63. Beecher, pp. 194, 195, 198, 199.
64. Chase, pp. 67–74.
65. "Homemade Wines," *Godey's Magazine* 62, Jan.-Jun. 1861, p. 459.
66. Porcher, pp. 195, 279, 253, 385–387, 507, 552.
67. "Persimmon Beer," *Southern Cultivator* 19, No. 11, Nov. 1861, p. 291.
68. S. Rutledge, *The Carolina Housewife* (Charleston, Babcock and Company, 1847), p. 83.
69. Randolph, p. 177.
70. Beecher, pp. 146–147.
71. "Pickling in Vinegar," *Godey's Lady's Book and Magazine* 62, 1861, p. 168; "Pickles," *Godey's Lady's Book and Magazine* 65, 1862, p. 500.
72. "Observations on Pickles," *Godey's Lady's Book and Magazine* 65, 1862, p. 395.
73. Olmsted, pp. 108–110.
74. *Ibid.*, p. 417.
75. H.C. Covey and D. Eisnach, *What the Slaves Ate* (Santa Barbara, CA, Greenwood Press, 2009), pp. 51, 64; Federal Writers' Project, *American Slave Interviews: Arkansas Narratives*, Vol. 2, Part 6 (Washington, D.C., Works Progress Administration, 1941), p. 97 [Thomas Ruffin]; Federal Writers' Project, *American Slave Interviews: Georgia Narratives*, Vol. 4, Part 1 (Washington, D.C., Works Progress Administration, 1941), p. 285 [Benny Dillard].
76. Covey and Eisnach, p. 65.
77. P. Samford, "The Archaeology of African American Slavery and Material Culture," *The William and Mary Quarterly* [3rd Ser.] 53, No. 1, Jan. 1996, pp. 87–114, 107.
78. Covey and Eisnach, pp. 62–63, 66.
79. E.F. Frazier, "The Negro Slave Family," *The Journal of Negro History* 15, No. 2, Apr. 1930, pp. 198–259, p. 243.
80. K.F. Kiple and V.H. King, *Another Dimension to the Black Diaspora: Diet, Disease and Racism* (Cambridge, Cambridge University Press, 1981), p. 89.
81. Covey and Eisner, pp. 52–52; Federal Writers' Project, *American Slave Interviews: Arkansas Narratives*, Vol. 2, Part 2 (Washington, D.C., Works Progress Administration, 1941), p. 72 [Betty Curlett], p. 67 [Zenia Culp].
82. A.L. Young, "Cellars and African-American Slave Sites: New Data from an Upland South Plantation," *Midcontinental Journal of Archaeology* 22, No. 1, Spring 1997, pp. 100–101.
83. T.A. Singleton, "The Archaeology of Slavery in North America," *Annual Review of Anthropology* 24, 1995, p. 124.
84. Kiple and King, p. 124.
85. *Ibid.*, p. 90.
86. *Ibid.*, 126.
87. Porcher, pp. 541–542.
88. *Ibid.*, p. 397.
89. *Ibid.*, p. 549.
90. *Ibid.*, p. 551.
91. "Making Vinegar," *Southern Cultivator* 19, No. 6, June 1861, p. 199; Porcher, pp. 576–577.
92. "Remarks on the Acclimation of Plants," *Report of the Commissioner of Agriculture for the Year 1863* (Washington, D.C., Government Printing Office, 1863), p. 517.
93. S. Wolcott, "Brewing a New American Tea Industry," *Geographical Review* 102, No. 3, July 2012, p. 352; J. Sumner, pp. 175, 179.
94. "Remarks on the Acclimation of Plants," p. 517.
95. "A New Vegetable for the South," *Southern Cultivator* 19, No. 12, Dec. 1861, pp. 303–304.
96. "The Peruvian Ground Cherry," *Southern Cultivator* 19, No. 8, Aug. 1861, p. 245.

Chapter 5

1. A. Gray, *A Manual of the Botany of the Northern United States: From New England to Wisconsin and South to Ohio and Pennsylvania Inclusive (the Mosses and Liverworts by Wm. S. Sullivant,) Arranged According to the Natural System* (Boston, James Munroe, 1848), p. 26.
2. A. Gray, *A Manual of Botany of the Northern United States 4th ed.* (New York, Ivison, Phinney, Blakeman, and Company, 1863), p. 25.

3. T.D. Mitchell, *Materia Medica and Therapeutics* (Philadelphia, Grambo and Company, 1850), p. 586.
4. *Ibid.*, pp. 586–587.
5. "Castor Oil and Poppy," *Yorkville Enquirer* 9, No. 16, April 22, 1863, p. 1.
6. A. Berman, "The Thomsonian Movement and Its Relation to American Pharmacy and Medicine," *Bulletin of the History of Medicine* 25, No. 6, Nov.-Dec. 1951, pp. 522, 525.
7. A. Flexner, *Medical Education in the United States and Canada, A Report to the Carnegie Foundation for the Advancement of Teaching, Carnegie Foundation Bulletin* No. 4, 1910.
8. "Report," *Report of the Commissioner of Agriculture for the Year 1862* (Washington, D.C., Government Printing Office, 1863), p. 23
9. "Report," *Report of the Commissioner of Agriculture for the Year 1863* (Washington, D.C., Government Printing Office, 1863), p. 6.
10. G.W. Hasegawa, "Pharmacy in the Civil War," *Pharmacy in History* 42, Nos. 3 and 4, 2000, p. 70.
11. *Confederate Receipt Book: A Compilation of Over Hundred Receipts Adapted to the Times* (Richmond, West and Johnson, 1863), pp. 13–14.
12. *Ibid.*, p. 14.
13. *General Directions for Collecting and Drying Medicinal Substances of the Vegetable Kingdom. List and descriptions of indigenous plants, etc., their medicinal properties, forms of administration and doses* (Richmond, Virginia, Confederate States of America, Surgeon General's Office, 1862), Documenting the American South, https://docsouth.unc.edu.
14. W.T. Grant, "Indigenous Medicinal Plants," *Confederate States Surgical and Medical Journal* 1, No. 6, June 1864, pp. 84–85.
15. J.H. Claiborne, "On the Use of *Phytolacca Decandra* in Camp Itch," *Confederate States Surgical and Medical Journal* 1, No. 3, March 1864, p. 39.
16. F.P. Porcher, *Resources of the Southern Fields and Forests* (Richmond, Evans and Cogswell, 1863).
17. G.B. Wood and F. Bache, *Dispensatory of the United States of America*, 6th ed. (Philadelphia, Grigg and Elliott, 1845), Appendix. Drugs and Medicines Not Officinal I, pp. 1221–1305.
18. Porcher, pp. 7–8.
19. *Ibid.*, p. 44.
20. *Standard Supply Table of the Indigenous Remedies for Field Service and the Sick in General Hospitals* (Richmond, Virginia, Surgeon General's Office, 1863), Documenting the American South, https://docsouth.unc.edu, p. 1; Mitchell, pp. 342–343.
21. Porcher, p. 84.
22. *Ibid.*, pp. 275–280.
23. *Standard Supply Table of the Indigenous Remedies for Field Service and the Sick in General Hospitals*, p. 2.
24. G.B. Wood and F. Bache, pp. 422–425.
25. *Ibid.*, p. 425.
26. *Standard Supply Table of the Indigenous Remedies for Field Service and the Sick in General Hospitals*, p. 2.
27. *Confederate Receipt Book*, p. 13.
28. *Standard Supply Table of the Indigenous Remedies for Field Service and the Sick in General Hospitals*, pp. 2, 4.
29. *The War of the Rebellion: A Compilation of the Official Records of the Union And Confederate Armies* Ser. 4, Vol. 2, p. 442 (Washington, D.C., Government Printing Office, 1900).
30. "Castor Oil and Poppy," *Yorkville Enquirer* 9, No. 16, April 22, 1863, p. 1.
31. G.W. Hasegawa, "Pharmacy in the Civil War," *Pharmacy in History* 42, Nos. 3 and 4, 2000, p. 72.
32. *Standard Supply Table of the Indigenous Remedies for Field Service and the Sick in General Hospitals*, p. 1.
33. Porcher, p. 377.
34. *Ibid.*, pp. 533–534.
35. G.B. Wood and F. Bache, p. 128.
36. *Standard Supply Table of the Indigenous Remedies for Field Service and the Sick in General Hospitals*, p. 1.
37. G.B. Wood and F. Bache, p. 530.
38. Porcher, p. 48; *Standard Supply Table of the Indigenous Remedies for Field Service and the Sick in General Hospitals*, p. 3.
39. G.B. Wood and F. Bache, pp. 528, 530.
40. *Standard Supply Table of the Indigenous Remedies for Field Service and the Sick in General Hospitals*, p. 4; Porcher, pp. 32–33.
41. Porcher, p. 351; G.B. Wood and F. Bache, p. 641.
42. J. Sumner, *American Household Botany: A History of Useful Plants 1620–1900* (Portland, OR, Timber Press, 2004), p. 251.
43. *Standard Supply Table of the Indigenous Remedies for Field Service and the Sick in General Hospitals*, p. 4.
44. G.B. Wood and F. Bache, p. 661.
45. "Salep, Made of Sweet Potatoes," *Southern Cultivator* 19, No. 12, Dec. 1861, pp. 299–300.
46. "Rural Hygiene," *Southern Cultivator* 19, No. 4, May 1861, p. 143.
47. Mitchell, p. 283.
48. A.C. Post and W.H. Van Buren, "Report on Military Hygiene and Therapeutics," in United States Sanitary Commission, *Military Medical and Surgical Essays Prepared for the United States Sanitary Commission 1862–1864* (Washington, D.C., 1865), p. 8.
49. B. Szczygiel and R. Hewitt, "Nineteenth-Century Medical Landscapes: John H. Rauch, Frederick Law Olmsted, and the Search for Salubrity," *Bulletin of the History of Medicine* 74, No. 4, Winter 2000, p. 717.
50. "Farmers' Houses," *Report of the Commissioner of Agriculture for the Year 1863* (Washington, D.C., Government Printing Office, 1863), pp. 318, 328.
51. "Observations on Atmospheric Humidity," *Report of the Commissioner of Agriculture for the*

Year 1865 (Washington, D.C., Government Printing Office, 1866), p. 547.

52. F.L. Olmsted, *A Journey in the Seaboard Slave States; with Remarks on Their Economy* (London, Sampson Low, Son, and Company, 1856), p. 158.

53. *Ibid.*, p. 474.

54. *Ibid.*, p. 418.

55. M.C. Mitchell, "Health and the Medical Profession in the Lower South, 1845–1860," *The Journal of Southern History* 10, No. 4, Nov. 1944, p. 436.

56. J.H. Warner, "A Southern Medical Reform: The Meaning of the Antebellum Argument for Southern Medical Education," *Bulletin of the History of Medicine* 57, No. 3, Fall 1983, p. 368.

57. Porcher, pp. 95–96.

58. "No Use for Quinine," *Southern Cultivator* 20, Nos. 9 and 10, Sept. and Oct. 1862, p. 184.

59. J. Jones, "Indigenous remedies of the Southern Confederacy which may be employed in the treatment of malarial fever," *Southern Medical and Surgical Journal* 17, 1861, pp. 673–718, 753–787.

60. "On the External Application of Oil of Turpentine as a Substitute for Quinine in Intermittent Fever," *Confederate States Medical and Surgical Journal* 1, No. 1, Jan. 1864, pp. 7–8.

61. F.A. Michaux, *The North American Sylva*, Vol. 1 (Philadelphia, R.P. Smith, 1853), p. 121.

62. Porcher, p. 59.

63. N.H. Franke, "Pharmacy and Pharmacists in the Confederacy," *The Georgia Historical Quarterly* 38, No. 1, March 1954, p. 19.

64. R. Graziose, P. Rojas-Silva, *et al.*, "Antiparasitic Compounds from *Cornus florida* L. with Activities Against *Plasmodium falciparum* and *Leishmania tarentolae*," *Journal of Ethnopharmacology* 142, No. 2, 2012, pp. 456–461.

65. J.A. Duke, *Handbook of Medicinal Herbs* (Boca Raton, CRC Press, 1985), p. 188.

66. Porcher, p. 412.

67. M.B. Beck, "On *Eupatorium* and *Serpentaria*," *Confederate States Surgical and Medical Journal* 1, No. 9, Sept. 1864, pp. 137–138.

68. Porcher, pp. 147, 372, 419, 421, 484.

69. *Ibid.*, pp. 8–9.

70. Mitchell, p. 348.

71. "A Simple Salve for Soldiers' Feet in Marching," *Southern Cultivator* 19, No. 11, Nov. 1861, p. 296.

72. Mitchell, p. 674.

73. *Ibid.*

74. *Ibid.*, pp. 354–358.

75. *Ibid.*, p. 565.

76. *Ibid.*, pp. 83–86.

77. S.D. Gross, *A Manual of Military Surgery* (Richmond, Confederate States Army, Ayres and Wade, 1863), p. 97.

78. Gross, pp. 171–172.

79. Mitchell, p. 569.

80. *Ibid.*, pp. 572–573, 575.

81. I. Gelber, "The Addict and His Drugs," *The American Journal of Nursing* 63, No. 7, July 1963, p. 54.

82. P.W. Gates, *Agriculture and the Civil War* (New York, Alfred A. Knopf, 1965), pp. 98–99.

83. W. Diamond, "Imports of the Confederate Government from Europe and Mexico," *The Journal of Southern History* 6, No. 4, Nov. 1940, p. 496.

84. Gross, p. 20.

85. *Ibid.*, p. 7.

86. *Ibid.*, p. 12.

87. Porcher, p. 95.

88. *Ibid.*, pp. 97, 99, 107.

89. D.L. Cowen, "History of Pharmacy and the History of the South," *Apothecary's Cabinet* No. 6, Spring 2003, p. 3.

90. S.C. Kenny, "'A Dictate of Both Interest and Mercy'? Slave Hospitals in the Antebellum South," *Journal of the History of Medicine and Allied Sciences* 65, No. 1, Jan. 2010, pp. 8–9.

91. D.H. Boster, "An 'Epeleptick' Bondswoman: Fits, Slavery, and Power in the Antebellum South," *Bulletin of the History of Medicine* 83, No. 2, Summer 2009, pp. 277, 284, 295.

92. C.C. Yates, *Observations on the Epidemic Now Prevailing in the City of New-York* (New York, George P. Scott and Co., 1832).

93. W. Swaim, "A Treatise on the Alternative and Curative Virtues of Swaim's Panacea, and for Its Application to the Different Diseases of the Human System" (Philadelphia, John Bioren, 1833), p. 19.

94. Federal Writers' Project, *American Slave Interviews: South Carolina Narratives* (Washington, D.C., Works Progress Administration, 1941) Vol. 14, Part 1, p. 10 [Victoria Adams]; Federal Writers' Project, *American Slave Interviews: Georgia Narratives* (Washington, D.C., Works Progress Administration, 1941), Vol. 4, Part 1, p. 212 [Mary Colbert]; Federal Writers' Project, *American Slave Interviews: Georgia Narratives* (Washington, D.C., Works Progress Administration, 1941), Vol. 4, Part 2, p. 273, p. 274 [Emma Hurley], and many others.

95. *American Slave Interviews: Georgia Narratives* (Washington, D.C., Works Progress Administration, 1941), Vol. 4, Part 2, p. 273, p. 274 [Emma Hurley].

96. *American Slave Interviews: Georgia Narratives* (Washington, D.C., Works Progress Administration, 1941), Vol. 4, Part 3, pp. 91, 126 [William McWhorter, Harriet Miller]; *American Slave Interviews: Texas Narratives* (Washington, D.C., Works Progress Administration, 1941), Vol. 4, Part 4, p. 41 [Millie Ann Smith]; *American Slave Interviews: Missouri Narratives* (Washington, D.C., Works Progress Administration, 1941), Vol. 10, p. 321 [Gus Smith].

97. *American Slave Interviews: Texas Narratives* (Washington, D.C., Works Progress Administration, 1941), Vol. 4, Part 1, p. 242 [Harriet Collins] and others.

98. M.P. Leone and G. Fry, "Conjuring in the Big House Kitchen: An Interpretation of African American Belief Systems Based on the Uses of Archeology and Folklore Sources," *The Journal of*

American Folklore 112, No. 445, Summer 1999, p. 382.

99. J. Sumner, *The Natural History of Medicinal Plants* (Portland, OR, Timber Press, 2000), pp. 20–21.

100. L. Perrin, "Resisting Reproduction: Reconsidering Slave Contraception in the Old South," *Journal of American Studies* 35, No. 2, Part 2, August 2001, pp. 261, 267.

101. *Ibid.*, pp. 261–262.

102. H. Gutman, *The Black Family in Slavery and Freedom* (New York, Pantheon, 1976), pp. 80–82, note.

Chapter 6

1. F.S. Holmes, *The Southern Farmer and Market Gardener* (Charleston, Burges and James, 1842).

2. W.N. White, *Gardening for the South* (New York, C.M. Saxton and Company, 1856).

3. "Landscape Gardens," *Southern Cultivator* 20, Nos. 9 and 10, Sept. and Oct. 1862, p. 164.

4. "Our Flower Seeds," *Southern Cultivator* 19, No. 5, May 1861, p. 152.

5. F. Gentil and L. Liger, *The Retired Gard'ner* (London, J. Tonson, 1706).

6. F.L. Olmsted, *A Journey in the Back Country* (New York, Mason Brothers, 1860), pp. 19–20, 42, 169.

7. J.R. Cothran, *Gardens and Historic Plants of the Antebellum South* (Columbia, University of South Carolina Press, 2003), p. 11.

8. "The Progress of Horticulture," *The Magazine of Horticulture, Botany, and All Useful Discoveries and Improvements in Rural Affairs* 29 No. 1, Jan. 1863, p. 1.

9. M. Rion, *Ladies' Southern Florist* (Columbia, SC, Peter B. Glass, 1860), p. 111.

10. *Ibid.*, p. 125.

11. *Ibid.*, p. 99.

12. F.P. Porcher, *Resources of the Southern Fields and Forests* (Richmond, Evans and Cogswell, 1863), pp. 36–39.

13. G. Kidd, "Trees—Their Beauty and Arrangement in Your Garden," *American Cotton Planter and Soil of the South* 1, No. 8, 1857, p. 242.

14. A.J. Downing, *A Treatise on the Theory and Practice of Landscape Gardening* (New York, Wiley and Putnam, 1841), p. 196.

15. R. Nelson, "Evergreens for the South—No. 3," *Southern Cultivator* 14, No. 7, July 1856, p. 219.

16. *Prince's Descriptive Catalogue of Fruit and Ornamental Trees*, 1844–1845, p. 65.

17. P.J. Berckmans, "Evergreens for the South—Part II," *Southern Cultivator* 18, No. 2, Feb. 1860, p. 62.

18. "Preventative of Murrain," *Southern Cultivator* 30, No. 1, Jan. 1872, p. 62.

19. Rion, p. 122.

20. "Our Home Should Be Beautiful," *Southern Cultivator* 19, No. 7, July 1861, p. 215.

21. W.N. White, pp. 28–49.

22. *Ibid.*, p. 86.

23. "Hints for July and August," *Southern Cultivator*, vol. 20. No. 7 and 8, July and Aug. 1862, p. 131, pp. 129–131.

24. "Garden Seeds," *Southern Cultivator* 22, No. 2, Feb. 1864, p. 32.

25. "Prepare Your Soil in the Best Manner," *Southern Cultivator* 22, No. 2, Feb. 1864, p. 32.

26. "Farmers' Gardens," *Report of the Commissioner of Agriculture for the Year 1863* (Washington, D.C., Government Printing Office, 1863), pp. 337–365.

27. *Ibid.*, pp. 345–346.

28. J. Sumner, *American Household Botany: A History of Useful Plants 1620–1900* (Portland, OR, Timber Press, 2004), pp. 101–102.

29. "Celery—How to Raise in the South," *Southern Cultivator* 19, No. 7, July 1861, p. 222.

30. W.N. White, pp. 291, 298–299.

31. *Ibid.*, pp. 294–295.

32. *Ibid.*, pp. 295, 297.

33. *Ibid.*, pp. 191–192.

34. *Ibid.*, pp. 189–191, pp. 256–257.

35. "Make Your Home Beautiful!" *Southern Cultivator* 19, No. 4, April 1861, p. 123.

36. F.L. Olmsted, *A Journey in the Back Country* (New York, Mason Brothers, 1860), p. 394.

37. *Ibid.*, pp. 32–33.

38. M. Rion, *Ladies' Southern Florist* (Columbia, SC, Peter B. Glass, 1860).

39. *Ibid.*, p. 101.

40. *Ibid.*, pp. 64, 65, 95.

41. *Ibid.*, pp. 34–39.

42. *Ibid.*, pp. 85, 81.

43. *Ibid.*, p. 90.

44. "Manure for Flower Beds," *Southern Cultivator* 19, No. 10, Oct. 1861, p. 275.

45. "The Flower Garden," *Southern Cultivator* 19, No. 7, July 1861, p. 222.

46. "Farmers' Gardens," p. 363.

47. *Ibid.*, 1863.

48. *Ibid.*, p. 364.

49. H.W. Beecher, *Plain and Pleasant Talk About Fruits, Flowers, and Farming* (New York, Derby and Jackson, 1859), p. 117.

50. "Recipes for the Ladies," *Southern Cultivator* 19, No. 1, Jan. 1861, p. 19.

51. J.H. Walden, *Soil Culture: Containing a Comprehensive View of Agriculture, Horticulture, Pomology, Domestic Animals, Rural Economy, and Agricultural* Literature (New York, Robert Sears, 1858), p. 194.

52. C. Randolph, *The Parlor Gardener: A Treatise on the House Culture of Ornamental Plants* (Boston, J.E. Tilton, 1861).

53. *Ibid.*, p. 43.

54. "House Plants," *Report of the Commissioner of Agriculture for the Year 1863* (Washington, D.C., Government Printing Office, 1863), p. 367.

55. E.S. Sprague, Flowers for the Parlor and Garden (Boston, J.E. Tilton, 1863).

56. "The Fuchsia," *American Agriculturalist* 21, No. 10, Oct. 1862, p. 305.

57. J.L. Russell, "The Ferns as Cultivated

Plants," *The Magazine of Horticulture* 27, April 1861, pp. 173–180.

58. "In-Door Gardening," *The Magazine of Horticulture* 28, Sept. 1862, pp. 408–410.

59. J. Sumner, pp. 332–333.

60. "House Plants," p. 366.

61. "The Water Garden," *Godey's Ladies Book and Magazine* 65, July–December 1862, pp. 439–441; "Fruit Culture," *Godey's Ladies Book and Magazine* 65, July–December 1862, p. 94.

62. E. Parrish, *The Phantom Bouquet* (Philadelphia, J. B. Lippincott and Company, 1863).

63. C.E. Beecher and H.B. Stowe, *American Woman's Home* (New York, J.B. Ford, 1869), pp. 91–103.

64. *Ibid.*, p. 63.

65. J.H. Walden, p. 194.

66. "A Neat Greenhouse," *American Cotton Planter and Soil of the South* 1, No. 3 n.s., March 1857, p. 28.

67. J. Sumner, pp. 354–355.

68. D.L. Dix, *The Garland of Flora* (Boston, S.G. Goodrich, 1829).

69. S.J. Hale, *Flora's Interpreter, or The American Book of Flowers and Sentiments* (Boston, Marsh, Capen, and Lyon, 1832).

70. A.H.L. Phelps, *Botany for Beginners: An Introduction to Mrs. Lincoln's Lectures on Botany* (New York, F.J. Huntington and Company, 1833).

71. P.J. Hatch, "African-American Gardens at Monticello," www.monticello.org.

72. J.D.B. DeBow, "Plantation Management," *DeBow's Review, Agricultural, Commercial, Industrial Progress and Resources* 14 n.s., Feb. 1853, p. 178.

73. H.C. Covey and D. Eisnach, *What the Slaves Ate* (Santa Barbara, CA, Greenwood Press, 2009), pp. 74–77.

74. R. Westmacott, *African-American Gardens and Yards in the Rural South* (Knoxville, University of Tennessee Press, 1992), p. 103.

75. *Ibid.*, p. 80.

76. Federal Writers' Project, *American Slave Interviews: Arkansas Narratives* (Washington, D.C., Works Progress Administration, 1941), Vol. 2, Part 1, p. 63 [James Baker]; Federal Writers' Project, *American Slave Interviews: Georgia Narratives* (Washington, D.C., Works Progress Administration, 1941), Vol. 4, Part 2, p. 245 [Easter Huff]; Federal Writers' Project, *American Slave Interviews: Georgia Narratives* (Washington, D.C., Works Progress Administration, 1941), Vol. 4, Part 1, p. 155 [Julia Bunch], and others.

77. G. Kidd, "Trees—Their Beauty and Arrangement in Your Garden," *American Cotton Planter and Soil of the South* 1, No. 8, Aug. 1857, p. 242.

78. B.J. Heath and A. Bennett, "'The Little Spots Allow'd Them': The Archeological Study of African-American Yards," *Historical Archeology* 34, No. 2, 2000, p. 43.

79. Federal Writers' Project, *American Slave Interviews: Arkansas Narratives* (Washington, D.C., Works Progress Administration, 1941), Vol. 5, Part 1, p. 315 [Rachel Perkins]; Federal Writers' Project, *American Slave Interviews: Georgia Narratives* (Washington, D.C., Works Progress Administration, 1941), Vol. 4, Part 1, p. 155 [Julia Bunch], Federal Writers' Project, *American Slave Interviews: Alabama Narratives* (Washington, D.C., Works Progress Administration, 1941), Vol. 1, p. 251 [Sally Reynolds].

80. E. Genovese, *Roll, Jordan, Roll: The World the Slaves Made* (New York, Random House, 1972), p. 535.

81. J. Sumner, p. 129.

82. A. Gray, *A Manual of the Botany of the Northern United States: from New England to Wisconsin and South to Ohio and Pennsylvania Inclusive (the Mosses and Liverworts by Wm. S. Sullivant,) Arranged According to the Natural System* (Boston, James Munroe, 1848).

83. C. Darwin to A. Gray, June 5, 1861, Darwin Correspondence Project, www.darwinproject.ac.uk.

84. A. Gray to C. Darwin, Feb. 16, 1864, Darwin Correspondence Project, www.darwinproject.ac.uk.

85. C. Darwin to A. Gray, Apr. 16, 1866. Darwin Correspondence Project, www.darwinproject.ac.uk.

86. A. Gray, *How Plants Grow* (New York, American Book Company, 1858), p. 1.

87. *Ibid.*, p. 94.

88. *Peter Parley's Illustrations of the Vegetable Kingdom* (Boston, B.B. Mussey, 1840,), pp. lxii-lxiii.

89. *Ibid.*, p. 37.

90. A. Gray, *How Plants Grow*, p. 121; "Wild Flowers," *Report of the Commissioner of Agriculture for the Year 1862* (Washington, D.C., Government Printing Office, 1863), p. 157.

Chapter 7

1. "Cotton from India," *Southern Cultivator* 20, No. 2, Feb. 1862, p. 37.

2. "The Zostera Marina," *The Illustrated London News* XLI, Oct. 18, 1862, p. 401.

3. Poetry of the Lancashire Cotton Famine 1861–65, https://cottonfaminepoetry.exeter.ac.uk.

4. "Zipporah Silk Cotton Seed," *Southern Cultivator* 20, No. 2, Feb. 1862, p. 64.

5. A.L. Olmsted and P.W. Rhode, "Biological Innovation and Productivity Growth in the Antebellum Cotton Economy," *The Journal of Economic History* 68 No. 4, Dec. 2008, p. 1138; J.H. Moore, "Cotton Breeding in the Old South," *Agricultural History* 30, No. 3, July 1956, p. 100.

6. J.H. Moore, pp. 101–102.

7. *Ibid.*, p. 101.

8. A.L. Olmsted and P.W. Rhode, p. 1125.

9. "Cotton as an Antiseptic," *Southern Cultivator* 19, No. 8, Aug. 1861, p. 246.

10. "Cotton Shoe Thread," *Southern Cultivator* 20, Nos. 11 and 12, Nov.-Dec. 1862, p. 214.

11. Note, *Scientific American* 1008, No. 2, Jan. 10, 1863, p. 19.

12. T. Gray, "The South Still Gropes in Eli Whitney's Shadow," *The Georgia Historical Quarterly* 20, No. 4, Dec. 1936, p. 346.

13. "Eli Whitney's Patent for the Cotton Gin," National Archives and Records Administration, www.archives.gov.

14. C.S. Aiken, "The Evolution of Cotton Ginning in the Southeastern United States," *Geographical Review* 63, No. 2, Apr. 1973, p. 197.

15. *Ibid.*, p. 200.

16. D. Penningroth, "Writing Slavery's History," *OAH Magazine of History* 23, No. 2, Apr. 2009, p. 14.

17. *Ibid.*

18. L.M. Alcott, *Transcendental Wild Oats: And Excerpts from the Fruitlands Diary* (Boston, Harvard Common Press, 1975).

19. "Flax and Flax-Cotton," *Report of the Commissioner of Agriculture for the Year 1862* (Washington, D.C., Government Printing Office, 1863), pp. 115, 116, 120.

20. F.P. Porcher, *Resources of the Southern Fields and Forests* (Richmond, Evans and Cogswell, 1863), p. 89.

21. S.M. Brooks, *Civil War Medicine* (Springfield, IL, C.C. Thomas, 1966), p. 86.

22. W.A. Hammond, "Lint for Wounded Soldiers," *New York Times*, Sept. 1, 1862, p. 1, www.nytimes.com.

23. Porcher, p. 88.

24. *Ibid.*, p. 272.

25. "Hemp Culture," *Report of the Commissioner of Agriculture for the Year 1863* (Washington, D.C., Government Printing Office, 1863), p. 93.

26. "Confederate Hemp," *Charleston Daily Courier*, July 12, 1862, p. 2.

27. Porcher, pp. 530–531.

28. *Ibid.*, p. 272.

29. M.E. Massey, *Ersatz in the Confederacy: Shortages and Substitutes on the Southern Homefront* (Columbia, University of South Carolina Press, 1993), p. 112.

30. Note, *Memphis Daily Appeal*, Nov. 13, 1862, p. 1.

31. Federal Writers' Project, *American Slave Interviews: Alabama Narratives* (Washington, D.C., Works Progress Administration, 1941), Vol. 1, p. 54 [Henry Cheatham], p. 70 [Sara Colquitt]; p. 72 [Mandy Cosby]; p. 95 [Katherine Eppes]; p. 97 [Reuben Fitzpatrick]; p. 105 [Angie Garrett]; p. 128 [Esther Green]; p. 149 [Joseph Holmes], p. 172 [Jane], p. 187 [Lucindy Jurdon].

32. "Domestic Manufactures," *Southern Cultivator* 20, Nos. 11 and 12, Nov. and Dec. 1862, p. 205; "A New Manufacture," *Southern Cultivator* 20, nos. 11 and 12, Nov. and Dec. 1862, p. 224.

33. J.E. Cashin, "Trophies of War: Material Culture in the Civil War Era," *Journal of the Civil War Era* 1, No. 3, Sept. 2011, pp. 339, 344–347.

34. "Spinning Wheels," *The Southern Confederacy* (Athens, Georgia), Nov. 7, 1862, p. 2.

35. B.J. Derbes, "Prison Productions: Textiles and Other Military Supplies at the Louisiana State Penitentiary in the Civil War," *Louisiana History: The Journal of the Louisiana Historical Association* 55, No. 1, Winter 2014, p. 52.

36. J. Giesberg, "Waging War Their Own Way: Women and the Civil War in Pennsylvania," *Pennsylvania Legacies* 13, No. 1–2, June 2013, pp. 22–23.

37. "The Comfort Cloak," *Southern Cultivator* 20, Nos. 11 and 12, Nov. and Dec. 1862, p. 205.

38. M.R. Wilson, "The Extensive Side of Nineteenth-Century Military Economy: The Tent Industry in the Northern United States During the Civil War," *Enterprise & Society* 2, No. 2, June 2001, p. 302.

39. H.H. Sibley, *Conical Tent*, U.S. Patent 14,740, Apr. 22, 1856, U.S. Patent and Trademark Office.

40. C.H. Quenzel, "General Henry Hopkins Sibley: Military Inventor," *The Virginia Magazine of History and Biography* 65, No. 2, Apr. 1956, pp. 168–169.

41. Porcher, p. 500.

42. J.R. Garrison, "George Thompson Garrison's Shelter Tent," *Buildings & Landscapes: Journal of the Vernacular Architecture Forum* 20, No. 1, Spring 2013, p. 129.

43. R.S. Coddington, "Hidden Treasures: Inside the Only Museum in America Dedicated Solely to Civil War Soldier Images," *Military Images* 34, No. 1, Winter 2016, p. 22.

44. B.J. Derbes, "Prison Productions: Textiles and Other Military Supplies at the Louisiana State Penitentiary in the Civil War," *Louisiana History: The Journal of the Louisiana Historical Association* 55, No. 1, Winter 2014, p. 51.

45. "Valuable Receipts," *Scientific American* 1008, No. 2, Jan. 10, 1863, p. 20.

46. "Gun Cotton and Collodion," *Scientific American* 109, Dec. 12, 1857, pp. 13–14.

47. M.R. Gilchrist, "Disease and Infection in the American Civil War," *The American Biology Teacher* 60, No. 4, Apr. 1998, pp. 259–260.

48. H.H. Cunningham, *Doctors in Gray*, 2nd ed. (Baton Rouge, Louisiana State University Press, 1960), p. 232.

49. H.P. Byram, "An Essay on the Culture and Manufacture of Silk," *Annual Report of the Commissioner of Patents for the Year 1847* [House of Representatives, Ex. Doc. No. 54] (Washington, D.C., Wendell and Van Benthuysen, 1848), p. 441.

50. "Additional Observations on the Ailanthus Silk-Worm of China," *Report of the Commissioner of Agriculture for the Year 1862* (Washington, D.C., Government Printing Office, 1863), p. 393.

51. J. Sumner, *American Household Botany: A History of Useful Plants 1620–1900* (Portland, OR, Timber Press, 2004), p. 291.

52. S. Campion, "Wallpaper Newspapers of the American Civil War," *Journal of the American Institute for Conservation* 34, No. 2, Summer 1995, pp. 130, 132.

53. *Natchez Courier*, May 20, 1862, p. 1.

54. *Charleston Mercury*, Jan. 18, 1862, p. 2.

55. A.J. Valente, "Changes in Print Paper During the 19th Century," *Proceedings of the Charleston Library Conference,* 2010, http://dx.doi.org/10.5703/1288284314836, pp. 210–212.
56. "Maize-paper and Maize-cloth," *Report of the Commissioner of Agriculture for the Year 1863* (Washington, D.C., Government Printing Office, 1863), pp. 436–438.
57. Porcher, pp. 274, 305–308, 518–522, 544–545.
58. *Ibid.*, pp. 180–186.
59. *Ibid.*, p. 178.
60. J. Sumner, pp. 100–101.
61. Porcher, p. 78.
62. "Domestic Dyes," *Southern Cultivator* 20, Nos. 5 and 6, May and June 1862, pp. 108–109.
63. Porcher, pp. 80, 204.
64. "Domestic Dyes," *Southern Cultivator* 20, Nos. 5 and 6, May and June 1862, pp. 108–109.

Chapter 8

1. R.B. Outland III, *Tapping the Pines: The Naval Stores Industry in the American South* (Baton Rouge, Louisiana State University, 2004), pp. 5–6.
2. "Shelter and Protection of Orchards," *Report of the Commissioner of Agriculture for the Year 1862* (Washington, D.C., Government Printing Office, 1863), p. 149.
3. *Ibid.*
4. "American Forests: Their Destruction and Preservation," *Report of the Commissioner of Agriculture for the Year 1865* (Washington, D.C., Government Printing Office, 1866), pp. 219–220.
5. J. Sumner, *American Household Botany: A History of Useful Plants, 1620–1900* (Portland, OR, Timber Press, 2004), p. 279
6. M.E. Massey, *Ersatz in the Confederacy: Shortages and Substitutes on the Southern Homefront* (Columbia, University of South Carolina Press, 1993), p. 109.
7. F.P. Porcher, *Resources of the Southern Fields and Forests* (Richmond, Evans and Cogswell, 1863), p. 237.
8. *Ibid.*, p. 525.
9. F.L. Olmsted, *A Journey in the Seaboard Slave States; with Remarks on Their Economy* (London, Sampson Low, Son, and Company, 1856), pp. 383, 482.
10. J.H. Belter, *Improvement in the Method of Manufacturing Furniture,* U.S. Patent No. 19,405, Feb. 23, 1858. U.S. Patent and Trademark Office.
11. J.H. Belter, *Machinery for Sawing Arabesque Chairs,* U.S. Patent No. 5,208, July 31, 1847, U.S. Patent and Trademark Office.
12. C. Vincent, "John Henry Belter's Patent Parlour Furniture," *Furniture History* 3, 1967, pp. 92, 95.
13. P.W. Gates, *Agriculture and the Civil War* (New York, Alfred A. Knopf, 1965), p. 83.
14. Porcher, p, 263.
15. *Ibid.*, pp. 341–342, 498.
16. Massey, p. 109.
17. Porcher, p. 591.
18. J. Sumner, p. 58.
19. Federal Writers' Project, *American Slave Interviews: Arkansas Narratives* (Washington, D.C., Works Progress Administration, 1941), Vol. 2, Part 2, p. 83 [Betty Curlett].
20. Massey, p. 112.
21. "Lights: Practical Hints for Hard Times," *Southern Cultivator* 20, Nos. 11 and 12, Nov. and Dec. 1862, p. 203.
22. Porcher, pp. 122–123.
23. H.D. Allen, "The Paper Money of the Confederate States. With Historical Data," *The Numismatist* 31, Jan. 1918, p. 7.
24. "Osier Willows for Hedging, Basket Making. Etc.," *Southern Cultivator* 19, No. 3, March 1861, p. 78.
25. Porcher, pp. 267, 273, 339, 340.
26. J.R. Garrison, "George Thompson Garrison's Shelter Tent," *Buildings & Landscapes: Journal of the Vernacular Architecture Forum* 20, No. 1, Spring 2013, p. 131.
27. "Wooden Shoes," *Southern Cultivator* 20, Nos. 7 and 8, July and Aug. 1862, p. 146.
28. B.M. Jordan, "'Our Work Is Not Yet Finished': Union Veterans and Their Unending Civil War, 1865–1872," *Journal of the Civil War Era* 5, No. 4, Dec. 2015, p. 490.
29. L.M. Alcott, *Hospital Sketches* [reprint of the 1863 edition] (Boston, Applewood Books, 1986), p. 62.
30. B.W. Jewett, *Artificial Leg,* U.S. Patent, 16,360, January 6, 1857, U.S. Patent and Trademark Office.
31. E.F. Palmer, *Improvement in Artificial Legs,* U.S. Patent 137,711, Apr. 8, 1873, U.S. Patent and Trademark Office.
32. H.G. Reinhardt, *Improvement in Artificial Legs,* U.S. Patent 41,535, Feb. 9, 1864, U.S. Patent and Trademark Office.
33. L. Tassius, *Improved Artificial Leg,* U.S. Patent 81,033, Aug. 11, 1868, U.S. Patent and Trademark Office.
34. L. Tassius, U.S. Patent 81,033.
35. T. Uren, *Improvement in Artificial Arms and Hands,* U.S. Patent 46,158, Jan. 1, 1865, U.S. Patent and Trademark Office.
36. G.A. Coco, *A Strange and Blighted Land—Gettysburg, the Aftermath of a Battle* (Gettysburg, Thomas Publications, 1995), p. 64.
37. D.G. Faust, *This Republic of Suffering: Death and the American Civil War* (New York, Alfred A. Knopf, 2008) p. 79.
38. *Ibid.*, p. 77.
39. C. Goodyear, *The Applications and Uses of Vulcanized Gum-Elastic, with Descriptions and Directions for Manufacturing Purposes* (New Haven, 1853), pp. 129–64.
40. C. Goodyear, *The Applications and Uses of Vulcanized Gum-Elastic, with Descriptions and Directions for Manufacturing Purposes* (New Haven, 1853) p. 133.
41. "Price list for The Union India Rubber Company, New York, August 1861,"

Ephemera Collection, Historic New England, www.historicnewengland.org.

42. Goodyear, *The Applications and Uses of Vulcanized Gum-Elastic, with Descriptions and Directions for Manufacturing* Purposes, pp. 28–29.

43. Olmsted, *A Journey in the Seaboard Slave States*, pp. 338–51.

44. *Ibid.*, p. 344.

45. P. Perry, "The Naval-Stores Industry in the Old South, 1790–1860," *The Journal of Southern History*, 34, No. 4, Nov. 1968, p. 514.

46. Olmsted, *A Journey in the Seaboard Slave States*, p. 345.

47. *Ibid.*, p. 347.

48. Outland, p. 6.

49. J.C. Duane, *Manual for Engineer Troops* (New York, D. Van Nostrand, 1862 and later editions).

50. T.F. Army, *Engineering Victory: How Technology Won the Civil War* (Baltimore: Johns Hopkins University Press, 2016), p. 196.

51. *Ibid.*, p. 35.

52. Duane, *Manual for Engineer Troops.*

53. U.S. Grant, *Memoirs of General U.S. Grant, Complete, Vol. 2*, Chapter LXIV (New York, Charles L. Webster, 1885).

54. Quoted by T.F. Army in *Engineering Victory: How Technology Won the Civil War*, p. 1.

55. Army, pp. 278–279.

56. D.H. Mahan, *A Treatise on Field Fortification*, 4th ed. (Richmond, VA, West and Johnson, 1862), p. 38.

57. *Ibid.*, p. 36.

58. J.C. Duane, *Manual for Engineer Troops*, 3rd ed. (New York, D. Van Nostrand, 1864), pp. 80–83.

59. Mahan, p. 45.

60. Duane, 1864, pp. 78–79.

61. *The War of the Rebellion: A Compilation of the Official Records of the Union and Confederate Armies*, vol. XXIV, part 2, Reports [May 16–August 10, 1863] (Washington, D.C., Government Printing Office, 1889), pp. 172, 180.

62. L.M. Brady, *War upon the Land: Military Strategy and the Transformation of Southern Landscapes during the Civil War* (Athens, University of Georgia Press, 2012), p. 122.

63. *Ibid.*, pp. 31–40.

64. "Virginia: Her Past, Present, and Future," *Report of the Commissioner of Agriculture for the Year 1864* (Washington, D.C., Government Printing Office, 1865), p. 20.

65. "American Forests: Their Destruction and Preservation," *Report of the Commissioner of Agriculture for the Year 1865* (Washington, D.C., Government Printing Office, 1866), pp. 215–216.

66. *Ibid.*, p. 218.

Afterword

1. *The Commemorative Celebration Held at Yale College* (New Haven, Tuttle, Morehouse, and Taylor, 1866), p. 16.

Bibliography

"Additional Observations on the Ailanthus Silk-Worm of China." *Report of the Commissioner of Agriculture for the Year 1862.* Washington, D.C., Government Printing Office, 1863, pp. 390–394.

"Agricultural Science and Literature." *The American Cotton Planter* 4 No. 3 n.s., March 1860, p. 107.

"Agriculture in Schools." *The Country Gentleman* 15 No. 6, February 9, 1860, p. 99.

Aiken, C.S. "The Evolution of Cotton Ginning in the Southeastern United States." *Geographical Review* 63 No. 2, April 1973, pp. 196–224.

Alcott, L.M. *Hospital Sketches* [reprint of the 1863 edition]. Boston, Applewood Books, 1986.

Alcott, L.M. *Transcendental Wild Oats: And Excerpts from the Fruitlands Diary.* Boston, Harvard Common Press, 1975.

Allen, B.T. *Sugaring Off: The Maple Sugar Paintings of Eastman Allen.* Williamstown, MA, Clark Art Institute, 2005.

Allen, H.D. "The Paper Money of the Confederate States. With Historical Data." *The Numismatist* 31, January 1918.

"American Forests: Their Destruction and Preservation." *Report of the Commissioner of Agriculture for the Year 1865.* Washington, D.C., Government Printing Office, 1866, pp. 219–220.

"Apples," *The Ladies' Home Magazine* 15, December 1860, p. 372.

Army, T.F. *Engineering Victory: How Technology Won the Civil War,* Baltimore, Johns Hopkins University Press, 2016.

"Artificial Guano. A Cheap and Good Fertilizer." *Southern Cultivator* 19 No. 1, January 1861, pp. 19–20.

Bailey, L.H. *Manual of Cultivated Plants,* rev. ed. New York, Macmillan, 1949.

Bandy, W.C. "Cotton Seed Manure." *The American Cotton Planter* 1 No. 3 n.s., March 1857, p. 326.

Barlow, C. "Anachronistic Fruits." *Arnoldia* 61, No. 2, 2001, pp. 14–21.

Beck, M.B. "On *Eupatorium* and *Serpentaria.*" *Confederate States Surgical and Medical* Journal 1 No. 9, September 1864, pp. 137–138.

Beecher, C. *Miss Beecher's Domestic Receipt-Book,* 3rd ed. New York, Harper and Brothers, 1862.

Beecher, C.E. and H.B. Stowe, *The American Woman's Home.* New York, J.B. Ford and Company, 1869.

Beecher, H.W. *Plain and Pleasant Talk about Fruits, Flowers, and Farming.* New York, Derby and Jackson, 1859.

Belter, J.H. *Improvement in the Method of Manufacturing Furniture,* U.S. Patent No. 19,405, February. 23, 1858, U.S. Patent and Trademark Office.

Belter, J.H. *Machinery for Sawing Arabesque Chairs,* U.S. Patent No. 5,208, July 31, 1847, U.S. Patent and Trademark Office.

Berckmans, P.J. "Evergreens for the South—Part II." *Southern Cultivator* 18 No. 2, February. 1860, p. 62.

Berman, A. "The Thomsonian Movement and Its Relation to American Pharmacy and Medicine." *Bulletin of the History of Medicine* 25 No. 6, November.-December 1951, pp. 519–538.

Boster, D.H. "An 'Epeleptick' Bondswoman: Fits, Slavery, and Power in the Antebellum South." *Bulletin of the History of Medicine* 83, No. 2, Summer 2009, pp. 271–301.

Brady, L.M. *War upon the Land: Military Strategy and the Transformation of Southern Landscapes during the Civil War.* Athens, University of Georgia Press, 2012.

"Bread for the South." *Southern Cultivator* 19 No. 1, January 1861, p. 13.

Brooks, S.M. *Civil War Medicine.* Springfield, IL, C.C. Thomas, 1966.

Burr, F., Jr. *The Field and Garden Vegetables of America.* Boston. J.E. Tilton and Company, 1865.

Byram, H.P. "An Essay on the Culture and Manufacture of Silk." *Annual Report of the Commissioner of Patents for the Year 1847 [House of Representatives, Ex. Doc. No. 54].* Washington, D.C., Wendell and Van Benthuysen, 1848, pp. 440–452.

"Campaigning Axioms." *Southern Cultivator* 19 No. 5, May, 1861, p. 194.

Campion, S. "Wallpaper Newspapers of the American Civil War." *Journal of the American Institute for Conservation* 34 No. 2, Summer 1995, pp. 129–140.

Cashin, J.E. "Trophies of War: Material Culture in the Civil War Era." *Journal of the Civil War Era* 1 No. 3, September. 2011, pp. 339–367.

"Castor Oil and Poppy." *Yorkville Enquirer* 9. No. 16, April 22, 1863, p. 1.
"Celery—How to Raise in the South." *Southern Cultivator* 19 No. 7, July 1861, p. 222.
Chase, A.W. *Dr. Chase's Recipes; or Information for Everybody: An Invaluable Collection of about 800 Recipes*, 30th ed. Ann Arbor, 1866.
"A Cheap and Ready Mode of Composting Night Soil." *The Country Gentleman* 16 No. 6, February 9, 1860, pp. 90–91.
Claiborne, J.H. "On the Use of *Phytolacca Decandra* in Camp Itch." *Confederate States Surgical and Medical Journal* 1 No. 3, March 1864, p. 39.
"Coal Oil and Cotton Seed Oil and Cake." *The Country Gentleman* 15 No. 1, January. 5, 1860, p. 19.
Coco, G.A. *A Strange and Blighted Land—Gettysburg, the Aftermath of a* Battle. Gettysburg, Thomas Publications, 1995.
Coddington, R.S. "Hidden Treasures: Inside the Only Museum in America Dedicated Solely to Civil War Soldier Images." *Military Images* 34 No. 1, Winter 2016, p. 22.
"The Comfort Cloak." *Southern Cultivator* 20 Nos. 11 and 12, November and December 1862, p. 205.
The Commemorative Celebration Held at Yale College. New Haven, Tuttle, Morehouse, and Taylor, 1866.
"Confederate Hemp." *Charleston Daily Courier*, July 12, 1862, p. 2.
Confederate Receipt Book: A Compilation of Over Hundred Receipts Adapted to the Times. Richmond, West and Johnson, 1863.
"Corn Culture—Prize Essay." *Southern Cultivator* 19 No. 3, March 1861, pp. 75–76.
Cornelius, J.D. *Slave Missions and the Black Church in the Antebellum South*. Columbia, University of South Carolina Press, 1999.
Cothran, J.R. *Gardens and Historic Plants of the Antebellum South*. Columbia, University of South Carolina Press, 2003.
"Cotton as an Antiseptic." *Southern Cultivator* 19 No. 8, August 1861, p. 246.
"Cotton Culture." *Southern Cultivator* 20 Nos. 11 and 12, November and December. 1862, p. 207.
"Cotton from India." *Southern Cultivator* 20 No. 2, February 1862, p. 37.
"Cotton Seed Huller," *Southern Cultivator* 17 No. 12, December 1859, p. 376.
"Cotton Seed Oil and Cake." *Southern Cultivator* 19 No. 1, January 1861, pp. 11–12.
"Cotton Shoe Thread." *Southern Cultivator* 20 Nos. 11 and 12, November and December 1862, p. 214.
Coulter, E.M. "The Movement for Agricultural Reorganization in the Cotton South during the Civil War." *Agricultural History* 1, No. 1, January 1927, pp. 3–17.
Covey. H.C. and D. Eisnach, *What the Slaves Ate*. Santa Barbara, CA, Greenwood Press, 2009.
"Cow Peas." *Southern Cultivator* 21 Nos. 9 and 10, September and October 1863, p. 110.
Cowen, D.L. "History of Pharmacy and the History of the South." *Apothecary's Cabinet* No. 6, Spring 2003, p. 3.
"Cultivation of Potatoes." *Southern Cultivator* 19 No. 3, March. 1861, pp. 83–84.
"The Cultivation of Sorghum." *Southern Cultivator* 19 No. 6, June 1861, p. 197.
"Culture of the Ruta Baga." *The Country Gentleman* 15 No. 12, March 22, 1860, p. 187.
Cunningham, H.H. *Doctors in Gray*, 2nd ed. Baton Rouge, Louisiana State University Press, 1960.
Darwin, C. *Journal of researches into the natural history and geology of the countries visited during the voyage of H.M.S. Beagle round the world, under the command of Capt. Fitz Roy, R.A.*, 2nd ed. London, John Murray, 1845.
DeBow, J.D.B. "Plantation Management." *DeBow's Review, Agricultural, Commercial, Industrial Progress and Resources* 14 n.s., February 1853, p. 178.
Derbes, B.J. "Prison Productions: Textiles and Other Military Supplies at the Louisiana State Penitentiary in the Civil War." *Louisiana History: The Journal of the Louisiana Historical Association* 55 No. 1, Winter 2014, pp. 40–64.
Diamond, W. "Imports of the Confederate Government from Europe and Mexico." *The Journal of Southern History* 6, No. 4, November 1940, pp. 470–503.
Dix, D.L. *The Garland of Flora*. Boston, S.G. Goodrich, 1829.
"Domestic Dyes." *Southern Cultivator* 20 Nos. 5 and 6, May and June, 1862, pp. 108–109.
"Domestic Manufactures." *Southern Cultivator* 20 Nos. 11 and 12, November and December 1862, p. 205.
Downing, A.J. *A Treatise on the Theory and Practice of Landscape Gardening*. New York, Wiley and Putnam, 1841.
Duane, J.C. *Manual for Engineer Troops*. New York, D. Van Nostrand, 1862 and later editions.
Duane, J.C. *Manual for Engineer Troops*, 3rd. ed. New York, D. Van Nostrand, 1864.
Duke, J.A. *Handbook of Medicinal Herbs*. Boca Raton, CRC Press, 1985.
"Eli Whitney's Patent for the Cotton Gin." National Archives and Records Administration, www.archives.gov
"An Essay." *The American Cotton Planter* 1 No. 9, September 1853, pp. 269–270.
"Experiments on Alcoholic Fermentation and Its Causes." *Report of the Commissioner of Agriculture for the Year 1864*. Washington, D.C., Government Printing Office, 1865, pp. 519–532.
"Farmers' Gardens." *Report of the Commissioner of Agriculture for the Year 1863*. Washington, D.C., Government Printing Office, 1863, pp. 337–365.
"Farmers' Houses." *Report of the Commissioner of Agriculture for the Year 1863*. Washington, D.C., Government Printing Office, 1863, pp. 313–337.
Faust, D.G. "The Rhetoric and Ritual of Agriculture in Antebellum South Carolina." *The Journal of Southern History* 45, No. 4, November1979, pp. 541–568.
Faust, D.G. *This Republic of Suffering: Death and the American Civil War*. New York, Alfred A. Knopf, 2008.

Federal Writers' Project. *American Slave Interviews: Alabama Narratives* Vol. 1. Washington, D.C., Works Progress Administration, 1941.
Federal Writers' Project. *American Slave Interviews: Arkansas Narratives* Vol. 2 Part 1. Washington, D.C., Works Progress Administration, 1941.
Federal Writers' Project. *American Slave Interviews: Arkansas Narratives* Vol. 2, Part 2. Washington, D.C., Works Progress Administration, 1941.
Federal Writers' Project. *American Slave Interviews: Arkansas Narratives* Vol. 2, Part 6. Washington, D.C., Works Progress Administration, 1941.
Federal Writers' Project. *American Slave Interviews: Arkansas Narratives* Vol. 5 Part 1. Washington, D.C., Works Progress Administration, 1941.
Federal Writers' Project. *American Slave Interviews: Georgia Narratives* Vol. 4 Part 1. Washington, D.C., Works Progress Administration, 1941.
Federal Writers' Project. *American Slave Interviews: Georgia Narratives* Vol. 4 Part 2. Washington, D.C., Works Progress Administration, 1941.
Federal Writers' Project. *American Slave Interviews: Georgia Narratives* Vol. 4 Part 3. Washington, D.C., Works Progress Administration, 1941.
Federal Writers' Project. *American Slave Interviews: South Carolina Narratives* Vol. 14 Part 1 Washington, D.C., Works Progress Administration, 1941.
Federal Writers' Project. *American Slave Interviews: Texas Narratives* Vol. 4 Part 1. Washington, D.C., Works Progress Administration, 1941.
Federal Writers' Project. *American Slave Interviews: Texas Narratives* Vol. 4 Part 4. Washington, D.C., Works Progress Administration, 1941.
"Female Volunteers." *Southern Cultivator* 19 No. 5, May 1861, p. 151.
Filmore, M. "First Annual Message." December 2, 1850, The Presidency Project, www.presidency.ucsb.edu.
"Flax and Flax-Cotton." *Report of the Commissioner of Agriculture for the Year 1862,* Washington, D.C., Government Printing Office, 1863, pp. 113–123.
Flexner, A. *Medical Education in the United States and Canada, A Report to the Carnegie Foundation for the Advancement of Teaching.* Carnegie Foundation Bulletin No. 4. New York, The Carnegie Foundation, 1910.
"The Flower-Garden," *Southern Cultivator* 19 No. 7, July 1861, p. 222.
Franke, N.H. "Pharmacy and Pharmacists in the Confederacy." *The Georgia Historical Quarterly* 38 No. 1, March 1954, pp. 11–28.
Frazier, E.F. "The Negro Slave Family," *The Journal of Negro History* 15 No. 2, April 1930, pp. 198–259.
"Fruit Culture." *Godey's Ladies Book and Magazine* 65, July-December 1862, p. 94.
"The Fruit Orchard and Nursery." *The American Cotton Planter and Soil of the South* 14 No. 2, February 1860, p. 82.
Fruitland advertisement. *The American Cotton Planter and Soil of the* South 14 No. 2, February 1860, p. 83.
"Fruits—Re-Production of Varieties." *Southern Cultivator* 19 No. 2, February 1861, pp. 60–61.
"The Fuchsia." *American Agriculturalist* 21, No. 10, October 1862, p. 305.
"Garden Seeds." *Southern Cultivator* 22 No. 2, February 1864, p. 32.
"Gardening for Farmers." *Southern Cultivator* 19 No. 2, February 1861, p. 63.
Garrison, J.R. "George Thompson Garrison's Shelter Tent." *Buildings & Landscapes: Journal of the Vernacular Architecture Forum* 20 No. 1, Spring 2013, pp. 129–132.
Gates, P.W. *Agriculture and the Civil War.* New York, Alfred A. Knopf, 1965.
Gelber, I. "The Addict and His Drugs." *The American Journal of Nursing* 63 No. 7, July 1963, pp. 52–56.
General Directions for Collecting and Drying Medicinal Substances of the Vegetable Kingdom. List and descriptions of indigenous plants, etc., their medicinal properties, forms of administration and doses. Richmond, VA, Confederate States of America, Surgeon General's Office, 1862. Documenting the American South, https://docsouth.unc.edu
Genovese, E. *Roll Jordan, Roll: The World the Slaves Made.* New York, Random House, 1972.
Gentil, F. and L. Liger. *The Retired Gard'ner.* London, J. Tonson, 1706.
Giesberg, J. "Waging War Their Own Way: Women and the Civil War in Pennsylvania." *Pennsylvania Legacies* 13 Nos. 1–2, June 2013, pp. 16–27.
Gilchrist, M.R. "Disease and Infection in the American Civil War." *The American Biology Teacher* 60 No. 4, April 1998, pp. 258- 262.
"A Good Cheap Pie." *The American Cotton Planter and Soil of the South* 14 No. 6, June 1860, p. 263.
Goodyear, C. *The Applications and Uses of Vulcanized Gum-Elastic, with Descriptions and Directions for Manufacturing Purposes.* New Haven, 1853.
"Gradual Deterioration of the Soil." *The American Cotton Planter* 4 No. 7 n.s., July 1860, pp. 303–304.
Grant, U.S. *Memoirs of General U.S. Grant, Complete,* Vol. 2, Chapter LXIV. New York, Charles L. Webster, 1885.
Grant, W.T. "Indigenous Medicinal Plants." *Confederate States Surgical and Medical Journal* 1 No. 6, June 1864, pp. 84–85.
Gray, A. *A Manual of Botany of the Northern United States,* 4th ed. New York, Ivison, Phinney, Blakeman, and Company, 1863.
Gray, A. *Manual of the Botany of the Northern United States, Including the District East of the Mississippi and North of North Carolina and*

Tennessee, 5th ed. New York, Ivison, Taylor, Blakeman, and Company, 1867.

Gray, A. *A Manual of the Botany of the Northern United States: from New England to Wisconsin and South to Ohio and Pennsylvania Inclusive (the Mosses and Liverworts by Wm. S. Sullivant,) Arranged According to the Natural System.* Boston, James Munroe, 1848.

Gray, A. *How Plants Grow.* New York, American Book Company, 1858.

Gray, T. "The South Still Gropes in Eli Whitney's Shadow." *The Georgia Historical Quarterly* 20 No. 4, December. 1936, pp. 345–355.

Graziose, R., P. Rojas-Silva, *et al.* "Antiparasitic compounds from *Cornus florida* L. with activities against *Plasmodium falciparum* and *Leishmania tarentolae.*" *Journal of Ethnopharmacology* 142 No. 2, 2012, pp. 456–461.

Gross, S.D. *A Manual of Military Surgery.* Richmond, Confederate States Army, Ayres and Wade, 1863.

"Guano." *Southern Field and Fireside* 2 No. 16, September 6, 1860, p. 126.

"Gun Cotton and Collodion." *Scientific American* 109, December 12, 1857, pp. 13–14.

Gutman, H. *The Black Family in Slavery and Freedom.* New York, Pantheon, 1976, pp. 80–82, note.

Hale, S.J. *Flora's Interpreter, or The American Book of Flowers and Sentiments.* Boston, Marsh, Capen, and Lyon, *1832.*

Hammond, W.A. "Lint for Wounded Soldiers." *New York Times,* September 1, 1862, p. 1, www.nytimes.com.

Harden, R.R. "The Rot in Cotton." *The Farmers' Register* 5 No. 1, May 1837, pp. 22–27.

Harter, J. *Plants: 2460 Copyright-Free Illustrations of Flowers, Trees, Fruits, and Vegetables.* Mineola, New York, Dover Publications. 1998.

Hasegawa, G.W. "Pharmacy in the Civil War." *Pharmacy in History* 42 Nos. 3 and 4, 2000, pp. 67–86.

Hatch, P.J. "African-American Gardens at Monticello." www.monticello.org.

Heath, B.J. and A. Bennett. "'The Little Spots Allow'd Them': The Archeological Study of African-American Yards." *Historical Archeology* 34 No. 2, 2000, pp. 38–55.

"Hemp Culture." *Report of the Commissioner of Agriculture for the Year 1863.* Washington, D.C., Government Printing Office, 1863, pp. 91–95.

Hill, A.F. *Economic Botany: A Textbook of Useful Plants and Products.* New York, McGraw-Hill, 1952.

"Hints for July and August." *Southern Cultivator* 20. Nos. 7 and 8, July and August, 1862, pp. 129–131.

"Hints for the Season." *Southern Cultivator* 21 Nos. 7 and 8, July and August 1863, p. 93.

Holmes, F.S. *Southern Farmer and Market Gardener.* Charleston, Burges and James, 1842.

"Homemade Wines,"*Godey's Magazine* 62, January—June 1861, p. 459.

"House Plants." *Report of the Commissioner of Agriculture for the Year 1863.* Washington, D.C., Government Printing Office, 1863, pp. 366–381.

"How to Raise Early Tomatoes." *Southern Cultivator* 19 No. 1, January 1861, p. 30.

"In-Door Gardening." *The Magazine of Horticulture* 28, September 1862, pp. 408–410.

"Indian or Mixed Pickles." *Godey's Lady's Book and Magazine* 65, November 1862, p. 500.

"Industry of Southern Women." *Southern Cultivator* 22 No. 1, January 1864, p. 22.

"Jerusalem Artichoke." Southern Cultivator 22 No. 3, March 1864, p. 51.

Jewett, B.W. *Artificial Leg,* U.S. Patent 16,360, January 6, 1857, U.S. Patent and Trademark Office.

Jones, J. "Indigenous Remedies of the Southern Confederacy Which May Be Employed in the Treatment of Malarial Fever." *Southern Medical and Surgical Journal* 17, 1861, pp. 673–718, 753–787.

Jordan, B.M. "'Our Work Is Not Yet Finished': Union Veterans and Their Unending Civil War, 1865–1872." *Journal of the Civil War Era* 5 No. 4, December 2015, pp. 484–503.

Kenny, S.C. "A 'Dictate of Both Interest and Mercy?' Slave Hospitals in the Antebellum South." *Journal of the History of Medicine and Allied Sciences* 65 No. 1, January 2010, pp. 1–47.

Kidd, G. "Trees—Their Beauty and Arrangement in Your Garden." *American Cotton Planter and Soil of the South* 1 No. 8, August 1857, p. 242.

Kiple, K.F. and V.H. King. *Another Dimension to the Black Diaspora: Diet, Disease and Racism.* Cambridge, Cambridge University Press, 1981.

"Landscape Gardens." *Southern Cultivator* 20 Nos. 9 and 10, September and October 1862, p. 164.

"Leaf Mould—How to Prepare." *Southern Cultivator* 19 No. 1, January 1861, p. 13.

Leone, M.P. and G. Fry. "Conjuring in the Big House Kitchen: An Interpretation of African American Belief Systems Based on the Uses of Archeology and Folklore Sources." *The Journal of American Folklore* 112 No. 445, Summer 1999, pp. 372–403.

"A Lesson of the Times: The South Must Be Independent." *Southern Cultivator* 19 No. 1, January 1861, p. 15.

"Lights: Practical Hints for Hard Times." *Southern Cultivator* 20 Nos. 11 and 12, November and December 1862, p. 203.

"Lime as a Manure." *Southern Cultivator* 19 No. 2, February 1861, pp. 49–50.

Mahan, D.H. *A Treatise on Field Fortification,* 4th ed. Richmond, VA, West and Johnson, 1862.

"Maize-paper and Maize-cloth." *Report of the Commissioner of Agriculture for the Year 1863,* Washington, D.C., Government Printing Office, 1863, pp. 436–438.

"Make Provision Crops." *Southern Cultivator* 20 Nos. 11 and 12, November and December 1862, p. 213.

"Make Your Home Beautiful!" *Southern Cultivator* 19 No. 4, April 1861, p. 123.

"Making Vinegar." *Southern Cultivator* 19 No. 6, June 1861, p. 199.

"Manure for Flower Beds." *Southern Cultivator* 19 No. 10, October 1861, p. 275.

"Manures." *Southern Cultivator* 23 No. 2, February 1864, pp. 25–26.

"The Marblehead Drumhead Cabbages." *New England Farmer* 13 No. 4, April 1861, p. 192.

"Marl." *Southern Cultivator* 19 No. 5, May 1861, p. 160.

Massey, M.E. *Ersatz in the Confederacy: Shortages and Substitutes on the Southern Homefront.* Columbia, University of South Carolina Press, 1993.

Menard, R.R. "How Sugar and Tobacco Planters Built Their Industries and Raised an Empire." *Agricultural History* 81, No. 3, Summer 2007, pp. 309–332.

Michaux, F.A. *The North American Sylva,* Vol. 1. Philadelphia, R.P. Smith, 1853.

Mitchell, M.C. "Health and the Medical Profession in the Lower South, 1845–1860." *The Journal of Southern History* 10, No. 4, November 1944, pp. 424–446.

Mitchell, T.D. *Materia Medica and Therapeutics* (Philadelphia, Grambo and Company, 1850.

Moore, J.H. "Cotton Breeding in the Old South." *Agricultural History* 30 No. 3, July 1956, pp. 95–104.

"A Neat Greenhouse." *American Cotton Planter and Soil of the South* 1 No. 3 n.s., March 1857, p. 28.

Nelson, R. "Evergreens for the South—No. 3." *Southern Cultivator* 14, No. 7, July 1856, p. 219.

"New Gas Works." *Scientific American* 13 No. 27, March 13, 1858, p. 212.

"A New Manufacture." *Southern Cultivator* 20 Nos. 11 and 12, November and December 1862, p. 224.

"A New Vegetable for the South." *Southern Cultivator* 19 No. 12, December 1861, pp. 303–304.

"No Use for Quinine." *Southern Cultivator* 20 Nos. 9 and 10, September and October 1862, p. 184

"Note." *Memphis Daily Appeal,* November 13, 1862, p. 1.

"Note." *Scientific American* 1008 No. 2, January 10, 1863, p. 19.

"Observations on Atmospheric Humidity." *Report of the Commissioner of Agriculture for the Year 1865.* Washington, D.C., Government Printing Office, 1866, pp. 520–550.

"Observations on Pickles." *Godey's Lady's Book and Magazine* 65, 1862, p. 395.

Olmsted, A.L. and P.W. Rhode. "Biological Innovation and Productivity Growth in the Antebellum Cotton Economy." *The Journal of Economic History* 68 No. 4, December 2008, pp. 1123–1171.

Olmsted, F.L. *A Journey in the Back Country.* New York, Mason Brothers, 1860.

Olmsted, F.L. *A Journey in the Seaboard Slave States; with Remarks on Their Economy.* London, Sampson Low, Son, and Company, 1856.

Olmsted, F.L. The *Cotton Kingdom: A Traveller's Observations on Cotton and Slavery in the American Slave States.* New York, Mason Brothers, 1862.

"On the External Application of Oil of Turpentine as a Substitute for Quinine in Intermittent Fever." *Confederate States Medical and Surgical Journal* 1 No. 1, January. 1864, pp. 7–8.

"The Orchard." *The American Cotton Planter and Soil of the South* 13 No. 1, January 1859, p. 26.

"Osier Willows for Hedging, Basket Making. Etc." *Southern Cultivator* 19 No. 3, March 1861, p. 78.

Ouchley, K. *Flora and Fauna of the Civil War: An Environmental Reference Guide.* Baton Rouge, Louisiana State University Press, 2010.

"Our Flower Seeds." *Southern Cultivator* 19 No. 5, May 1861, p. 152.

"Our Home Should Be Beautiful." *Southern Cultivator* 19 No. 7, July 1861, p. 215.

Outland R.B. III. *Tapping the Pines: The Naval Stores Industry in the American South.* Baton Rouge, Louisiana State University, 2004.

Palmer, E.F. *Improvement in Artificial Legs,* U.S. Patent 137,711, April 8, 1873, U.S. Patent and Trademark Office.

Parrish, E. *The Phantom Bouquet.* Philadelphia, J.B. Lippincott and Company, 1863.

"Peas Beans and Mangolds." *The Country Gentleman* 15 No. 21, May 24, 1860, p. 333.

Penningroth, D. "Writing Slavery's History." *OAH Magazine of History* 23 No. 2, April 2009, pp. 13–20.

Perrin, L. *"Resisting Reproduction: Reconsidering Slave Contraception in the Old South." Journal of American Studies 35 No. 2 Part 2, August 2001, pp. 255–274.*

Perry, P. "The Naval-Stores Industry in the Old South, 1790–1860." *The Journal of Southern History* 34 No. 4, November 1968, pp. 509–526.

"Persimmon Beer." *Southern Cultivator* 19 No. 11, November 1861, p. 291.

"The Peruvian Ground Cherry." *Southern Cultivator* 19 No. 8, August 1861, p. 245.

Peter Parley's Illustrations of the Vegetable Kingdom. Boston, B.B. Mussey, 1840.

Petrakis, P.V., K. Spanos, A. Feest, and E. Daskalakou. "Phenols in Leaves and Bark of *Fagus sylvatica* as Determinants of Insect Occurrences." *International Journal of Molecular Sciences* 12, 2011, 2769–2782.

Phelps, A.H.L. *Botany for Beginners: An Introduction to Mrs. Lincoln's Lectures on Botany.* New York, F.J. Huntington and Company, 1833.

"Pickles." *Godey's Lady's Book and Magazine* 65, 1862, p. 500.

"Pickling in Vinegar." *Godey's Lady's Book and Magazine* 62, 1861, p. 168.

"Plant Corn!" *Southern Cultivator* 19 No. 3, March 1861, pp. 84–85.

"Plantation Work for June." *The American Cotton Planter* 1 No. 3 n.s., March 1857, p. 163.

Poetry of the Lancashire Cotton Famine 1861–65. https://cottonfaminepoetry.exeter.ac.uk.

Porcher, F.P. *Resources of the Southern Fields and Forests.* Richmond, Evans and Cogswell, 1863.

Post, A.C. and W.H. Van Buren. "Report on Military Hygiene and Therapeutics," in United States

Sanitary Commission, *Military Medical and Surgical Essays Prepared for the United States Sanitary Commission 1862–1864*. Washington, D.C., 1865, pp. 1–27.
"Potash—A Hint to Farmers." *Southern Cultivator* 19 No. 12, December 1861 p. 305.
Powell, R.D. "Agriculture." *The American Cotton Planter* 1 No. 3 n.s., March 1857, pp. 326–328.
"Premium List of the First Annual Fair of the Alabama State Agricultural Society." *The American Cotton Planter* 3 No. 6, June 1855, pp. 168–176.
"Prepare Your Soil in the Best Manner." *Southern Cultivator* 22 No. 2, February 1864, p. 32.
"Preventative of Murrain." *Southern Cultivator* 30 No. 1, January 1872, p. 62.
"Price list for The Union India Rubber Company." New York, August 1861, Ephemera Collection, Historic New England www.historicnewengland.org.
Prince's Descriptive Catalogue of Fruit and Ornamental Trees. Flushing, NY, William R. Prince and Company, 1844–1845.
"Prize Essay on Commercial Manures." *Southern Cultivator* 19 No. 2, February 1861, pp. 42–43.
"The Progress of Horticulture." *The Magazine of Horticulture, Botany, and All Useful Discoveries and Improvements in Rural Affairs* 29 No. 1, January 1863, p. 1.
"Queen Victoria's Children." *Southern Cultivator* 19 No. 2, February 1861, p. 59.
Quenzel, C.H. "General Henry Hopkins Sibley: Military Inventor." *The Virginia Magazine of History and Biography* 65 No. 2, April 1956, pp. 166–176.
Radford, A.E., H.E. Ahles, and C.R. Bell. *Manual of the Vascular Flora of the Carolinas.* Chapel Hill, University of North Carolina Press, 1968.
Randolph, C. *The Parlor Gardener: A Treatise on the House Culture of Ornamental Plants.* Boston, J.E. Tilton, 1861.
Randolph, M. *The Virginia House-Wife* [reprint of the 1824 edition]. Columbia, University of South Carolina Press, 1984.
Ravenel, H.W. *Private Journal, 1865–1866,* September, 1865, p. 59. University of South Carolina University Libraries https://digital.tcl.sc.edu/digital/collection.
"Recipes for the Ladies." *Southern Cultivator* 19 No. 1, January 1861, p. 19.
"Refuse of Sorgho." *Southern Cultivator* 21 Nos. 7 and 8, July and August 1863, p. 94.
Reinhardt, H.G. *Improvement in Artificial Legs,* U.S. Patent 41,535, February 9, 1864, U.S. Patent and Trademark Office.
"Remarks on the Acclimation of Plants." *Report of the Commissioner of Agriculture for the Year 1863.* Washington, D.C., Government Printing Office, 1863), pp. 512–517.
"Report." *Report of the Commissioner of Agriculture for the Year 1863.* Washington, D.C., Government Printing Office, 1863, pp. 4–25.
"Report and Review." *American Cotton Planter* 2 No. 3, March 1854, pp. 233–235.
Rice, E.L. "Allelopathic Effects of Andropogon virginicus and Its Persistence in Old Fields." *American Journal of Botany* 59 No. 7, pp. 752–755.
Rion, M. *Ladies' Southern Florist.* Columbia, SC, Peter B. Glass, 1860.
"Rotation for a Small Farm." *The Country Gentleman* 15 No. 1, January 5, 1860, p. 11.
"Rural Hygiene." *Southern Cultivator* 19 No. 4, May 1861, p. 143.
Russell, J.L. "The Ferns as Cultivated Plants." *The Magazine of Horticulture* 27, April 1861, pp. 173–180.
Rutledge, S. *The Carolina Housewife.* Charleston, Babcock and Company, 1847.
"Salep, Made of Sweet Potatoes." *Southern Cultivator* 19 No. 12, December 1861, pp. 299–300.
Samford, P. "The Archaeology of African American Slavery and Material Culture." *The William and Mary Quarterly* [3rd Ser.] 53, No. 1, January 1996, pp. 87–114.
Sanderson, E. "Feeding Hogs on Cotton Seed." *The American Cotton Planter* 1 No. 3 n.s., March 1857, p. 315.
Sanderson, J.M. *Camp Fires and Camp Cooking: Culinary Hints for the Soldier.* Washington, D.C., Government Printing Office, 1862.
"Saving Manure." *Southern Cultivator* 19 No. 7, July 1861, p. 231.
"Shall We Raise Tobacco?" *The New England Farmer* 15 No. 12, December 1863, pp. 377–378.
"Shall We Raise Tobacco?" *The New England Farmer* 16 No. 1, January 1864, pp. 24–25.
"Shelter and Protection of Orchards." *Report of the Commissioner of Agriculture for the Year 1862.* Washington, D.C., Government Printing Office, 1863, pp. 147–155.
Sibley, H.H. *Conical Tent,* U.S. Patent 14,740, April 22, 1856, U.S. Patent and Trademark Office.
"A Simple Salve for Soldiers' Feet in Marching." *Southern Cultivator* 19 No. 11, November 1861, p. 296.
Singleton, T.A. "The Archaeology of Slavery in North America." *Annual Review of Anthropology* 24, 1995, pp. 119–140.
"Some Outlines of the Agriculture of Maine—Maple Sugar and Sirup." *Report of the Commissioner of Agriculture for the Year 1862.* Washington, D.C. Government Printing Office, 1863, pp. 39–59.
"Sorghum for Soiling." *The Country Gentleman* 15 No. 13, March 29, 1860, p. 205.
"Southern Native Trees and Shrubs." *The American Cotton Planter* 1, July 1853, p. 223.
"Spading and Subsoiling." *Southern Cultivator* 23 No. 2, February 1864, p. 28.
"Sparkling Wine—How Made!" *Southern Cultivator* 19 No. 2, February 1861, p. 63.
"Spinning Wheels." *The Southern Confederacy.* November 7, 1862, p. 2.
Sprague, E.S. *Flowers for the Parlor and Garden.* Boston, J.E. Tilton, 1863.
Standard Supply Table of the Indigenous Remedies for Field Service and the Sick in General Hospitals. Richmond, VA, Surgeon General's Office,

1863. Documenting the American South, https://docsouth.unc.edu
Stewart, M. A "From King Cane to King Cotton: Razing Cane in the Old South." *Environmental History* 12 No. 1, January 2007, pp. 59–79.
Sumner, J. *American Household Botany: A History of Useful Plants 1620–1900.* Portland, OR, Timber Press, 2004.
Sumner, J. *The Natural History of Medicinal Plants.* Portland, OR, Timber Press, 2000.
"Supplying Plant Food at the Surface." *The Country Gentleman* 15 No. 2, January 12, 1860, p. 27.
Swaim, W. *A Treatise on the Alternative and Curative Virtues of Swaim's Panacea, and for Its Application to the Different Diseases of the Human System.* Philadelphia, John Bioren, 1833.
Szczygiel, B. and R. Hewitt. "Nineteenth-Century Medical Landscapes: John H. Rauch, Frederick Law Olmsted, and the Search for Salubrity." *Bulletin of the History of Medicine* 74 No. 4, Winter 2000, pp. 708–734.
Tassius, L. *Improved Artificial Leg,* U.S. Patent 81,033, August 11, 1868, U.S. Patent and Trademark Office.
Taylor, P.S. "Plantation Laborer before the Civil War." *Agricultural History* 28 No. 1, January 1954, pp. 1–21.
Uren, T. *Improvement in Artificial Arms and Hands,* U.S. Patent 46,158, January 1, 1865, U.S. Patent and Trademark Office.
U.S. Centennial Commission, *International Exhibition, 1876 [Reports]* Vol. 9, Washington, U.S. Government Printing Office, 1876.
U.S. Quartermaster General. *Bread and Bread Making.* Washington, D.C., Government Printing Office, 1864.
Valente, A.J. "Changes in Print Paper During the 19th Century." *Proceedings of the Charleston Library Conference,* 2010. http://dx.doi.org/10.5703/1288284314836.
"Valuable Receipts." *Scientific American* 1008 No. 2, January 10, 1863, p. 20.
"Various Inquiries Answered." *The American Cotton Planter and Soil of the South* 3 No. 11, November 1859, p. 338.
Vincent, C. "John Henry Belter's Patent Parlour Furniture." *Furniture History* 3, 1967, pp. 92–99.
"Virginia: Her Past, Present, and Future." *Report of the Commissioner of Agriculture for the Year 1864.* Washington, D.C., Government Printing Office, 1865), pp. 16–42.
Walden, J.H. *Soil Culture: Containing a Comprehensive View of Agriculture, Horticulture, Pomology, Domestic Animals, Rural Economy, and Agricultural Literature.* New York, Robert Sears, 1858.
The War of the Rebellion: A Compilation of the Official Records of the Union and Confederate Armies, Vol. 2, Ser. 4. Washington, D.C., Government Printing Office, 1900.
The War of the Rebellion: a Compilation of the Official Records of the Union and Confederate Armies, Vol. XXIV, part 2, Reports [May 16–August 10, 1863]. Washington, D.C., Government Printing Office, 1889.
Warder, J.A. *Hedges and Evergreens.* New York, O.A. Moore, 1859.
Warner, J.H. "A Southern Medical Reform: The Meaning of the Antebellum Argument for Southern Medical Education." *Bulletin of the History of Medicine* 57 No. 3, Fall, 1983, pp. 364–381.
"The Water Garden." *Godey's Ladies Book and Magazine* 65, July-December 1862, pp. 439–441.
Westmacott, R. *African-American Gardens and Yards in the Rural South.* Knoxville, University of Tennessee Press, 1992.
"When to Sell Hay." *American Agriculturalist* 21 No. 10, October 1862, p. 301
White, W.N. *Gardening for the South* (New York, C.M. Saxton and Company, 1856).
"Wild Flowers." *Report of the Commissioner of Agriculture for the Year 1862.* Washington, D.C., Government Printing Office, 1863, pp. 155–162.
"The Willow and Osage Orange." *The American Cotton Planter* 3 No. 5, May 1855, p. 148.
Wilson, M.R. "The Extensive Side of Nineteenth-Century Military Economy: The Tent Industry in the Northern United States during the Civil War." *Enterprise & Society* 2 No. 2, June 2001, p. 302.
Wolcott, S. "Brewing a New American Tea Industry." *Geographical Review* 102 No. 3, July 2012, pp. 512–517.
"Women and Agriculture." *Southern Cultivator* 19 No. 4, April 1861, pp. 111–112.
Wood, G.B. and F. Bache, *Dispensatory of the United States of America,* 6th ed. Philadelphia, Grigg and Elliott, 1845, Appendix. Drugs and Medicines Not Officinal I, pp. 1221–1305.
"Wooden Shoes." *Southern Cultivator* 20 Nos. 7 and 8, July and August 1862, p. 146.
Works Progress Administration. *American Slave Interviews: Missouri Narratives* Vol. 10 Washington, D.C., Works Progress Administration, 1941.
Yates, C.C. *Observations on the Epidemic Now Prevailing in the City of New-York.* New York, George P. Scott and Company, 1832.
Young, A.L. "Cellars and African-American Slave Sites: New Data from an Upland South Plantation." *Midcontinental Journal of Archaeology* 22 No. 1, Spring, 1997, pp. 95–115.
"Zipporah Silk Cotton Seed." *Southern Cultivator* 20 No. 2, February 1862, p. 64.
"The Zostera Marina." The Illustrated London News XLI, October 18, 1862, p. 401.

Index